THE WAY OF THE HUMAN

THE QUANTUM
PSYCHOLOGY NOTEBOOKS

OTHER BOOKS BY STEPHEN WOLINSKY

Trances People Live
Healing Approaches in Quantum Psychology
ISBN 0-9626184-2-X, The Bramble Company

Quantum Consciousness
The Guide to Experiencing Quantum Psychology
ISBN 0-9626184-8-9, Bramble Books

The Tao of Chaos
Essence and the Enneagram
ISBN 1-883647-02-9, Bramble Books

The Dark Side of the Inner Child
The Next Step
ISBN 1-883647-00-2, Bramble Books

Hearts on Fire
The Tao of Meditation
ISBN 1-884997-25-2

THE WAY OF THE HUMAN

THE QUANTUM
PSYCHOLOGY NOTEBOOKS

VOLUME II

THE FALSE CORE
AND THE FALSE SELF

STEPHEN H. WOLINSKY, PH.D.

© 1999
All rights reserved.

For information write to:
Stephen H. Wolinsky, Ph.D.
Quantum Institute®
101 Grand Avenue, Suite 11
Capitola California 95010
(831) 464-0564

ISBN: 0-9670362-1-6

First printing 1999
Printed in Canada

THE AUTHOR

Stephen H. Wolinsky, Ph.D., began his clinical practice in Los Angeles, California in 1974. A Gestalt and Reichian therapist and trainer, he led workshops in Southern California. He was also trained in Classical Hypnosis, Psychosynthesis, Psychodrama/Psychomotor, and Transactional Analysis. In 1977 he journeyed to India, where he lived for almost six years studying meditation. He moved to New Mexico in 1982 to resume a clinical practice. There he began to train therapists in Ericksonian Hypnosis and family therapy. Dr. Wolinsky also conducted year-long trainings entitled: Integrating Hypnosis with Psychotherapy, and Integrating Hypnosis with Family Therapy. Dr. Wolinsky is the author of *Trances People Live: Healing Approaches in Quantum Psychology®, Quantum Consciousness: The Guide to Experiencing Quantum Psychology®, The Tao of Chaos: Quantum Consciousness Volume II, The Dark Side of the Inner Child* (Bramble Books) and *Hearts on Fire: The Roots of Quantum Psychology*. He is presently completing a three volume set entitled *The Way of the Human: The Quantum Psychology Notebooks*. He is the founder of Quantum Psychology®. Dr. Wolinsky presently resides in Capitola, California. He can be reached for workshop information by calling (831) 464-0564 or by FAX at (831) 479-8233.

DEDICATION

To the memory of Shri Nisargadatta Maharaj,
my Guru and teacher and
the grandfather of Quantum Psychology

To the memory of Alfred Korzybski
The father of General Semantics

To the memory of G.I. Gurdjieff
the Father of the Fourth Way

To Oscar Ichazo,
Father of the Modern Day Enneagram
which he calls the Enneagon

ACKNOWLEDGEMENTS

A special thank-you and acknowledgement to Dianne Postnieks, who inspired me to originate and create the False Core-False Self protocal questions.

Special thanks to Allen Horne for his editorial assistance.

Thanks to Susan Briley and Marylu Erlandson (Word processing).

And to my Divine Leni, for supporting this 5 1/2 year process.

The False-Core-False Self and the I-dentity work can be considered the preparatory or preliminary practise in realizing the purpose of Quantum Psychology. The discovery of **WHO YOU ARE** and **YOU ARE THAT QUANTUM CONSCIOUSNESS**.

Stephen H. Wolinsky

Quantum Psychology is an extension and continuation of Nisargadatta Maharaj—i.e., Advaita Vedanta—and it has several premises which form its basic core:

Advaita	Vedanta
There is only **THAT ONE SUBSTANCE**	Neti-Neti
Not two or more substances	(Sanskit for Not This-Not This)

SUMMARY OF NISARGADATTA MAHARAJ

1. There is only **ONE SUBSTANCE.**
2. What you know about yourself came from outside of you, therefore discard it.
3. Question everything, do not believe anything.
4. In order to find out who you are, you must first find out who you are not.
5. In order to let go of something, you must first know what it is.
6. The experien*cer* is contained within the experience itself.
7. Anything you think you are—you are NOT.
8. Hold onto the **I AM**, let go of everything else.
9. Anything you know about you can not be.

These nine principles form the core of
Quantum Psychology

Developmental Psychology and the Enneagram can be viewed as a map or model *only*. Maps can be utilized as vehicles to discover where the "I" you call yourself is identified.

The maps or models are very limited, and when we view the "world" or "our self" through these maps or models, or imagine we are these models, we see only what the map or model permits us to see through its lens.

Therefore, use the map—but realize it is a model, a lens to view reality.

It is not reality and does not describe everyone.

Then—give it up as soon as possible.

Believing in models as a lens will only yield more models and the repetition-compulsion that is contained within that lens/model itself.

This can be as dangerous to finding out **WHO YOU ARE** as believing in the False Core-False Self.

Stephen H. Wolinsky

TABLE OF CONTENTS

DEDICATION vi

INTRODUCTION xii

CHAPTER I:
THE FALSE CORE AND ITS BEGINNING 1

CHAPTER II:
THE FALSE CORE, AN OVERVIEW 51

CHAPTER III:
THE FALSE CORE 83

CHAPTER IV:
QUANTUM PSYCHOLOGY AT A FALSE CORE—
FALSE SELF WORKSHOP 99

CHAPTER V:
FUNDAMENTAL UNDERSTANDINGS REVISITED 115

CHAPTER VI:
THE WAY OF THE IMPERFECTIONIST 133

CHAPTER VII:
THE WAY OF THE WORTHLESS 149

CHAPTER VIII:
THE WAY OF THE NOT-DOER 197

CHAPTER IX
THE WAY OF THE INADEQUATE 237

CHAPTER X
THE WAY OF THE NON-EXISTENT 277

CHAPTER XI
THE WAY OF THE LONER 311

CHAPTER XII
THE WAY OF THE INCOMPLETE 335

CHAPTER XIII
THE WAY OF THE POWERLESS 349

CHAPTER XIV
THE WAY OF THE LOVELESS 363

CHAPTER XV
VOLUME II REVIEW 375

CHAPTER XVI
THE FALSE CORE AND ITS TRIGGERS 393

CHAPTER XVII
FORCE THEORY 403

CHAPTER XVIII
THE FALSE CORE KITE 409

APPENDIX:
CLOSING REMARKS 433

REFERENCES 439

INTRODUCTION

For many years, like children of the '60s, "I" was a seeker after truth. From drugs to sex, from psychotherapy to yoga, from Buddhism to the Sufis, from the Old World of Asia to the New Physics. By the 1980s, "I" had my answer to many questions. No longer was "I" troubled by **WHO AM "I?"** "I" knew. "I" was no longer suffering. "I" was myself. But as "I" plunged further and further into the blissful spaciousness of my own **ESSENCE** and beyond, "I" was confronted with what "I" imagined were pre-verbal structures which prevented me from being permanently established or stabilizing in what "I" have called Quantum Consciousness, the no-state state and beyond. It became clear that there were pitfalls—glitches in thinking and understanding which prevented "me" from becoming permanently established in being fully human and, simultaneously, in the underlying unity of *Quantum Consciousness*.

I decided to look within three models for direction: first, Developmental Psychology because I believed that it was the pre-verbal cognitions of an infant that distorted perception and unknowingly distracted one from remaining in the Quantum unity; second, the Enneagram, and third, Wilhelm Reich's character analysis.

Quantum Psychology further realized that there is one primary core concept we hold about ourselves and that this is the key to, and cornerstone of, our entire psychology. Quantum Psychology originally referred to this as the *Organizing Principle* in 1986; but since 1994 it has been called the False Core. It is a False Core because it represents a False Core concept or conclusion which organizes every thought, feeling, emotion, association, action, reaction, fantasy, etc. In short, everything "you" call "you" and which encompasses your entire psychology.

The Enneagram somehow suggests this understanding. Please note, this is not an Enneagram book, nor does Quantum Psychology agree with much of what is occurring within the Enneagram world. (See section—The Rise and Fall of the Enneagram.)

In *The Tao of Chaos*, I stressed that identities and beliefs are coverings or layers which appear to be created by the observer but which are actually part of the observer-identity complex. This False Core represents the organizing principle of our entire psychology or "individual mind" which shrinks and fixates awareness, thereby, preventing us from entering into our own **ESSENCE**, and ultimately, into the **VOID OF UNDIFFERENTIATED CONSCIOUSNESS**.

With this as the contextual understanding, "I" began looking at infantile understanding, and the beliefs and conclusions put forth by Developmental Psychology which are preconscious and pre-representational and underlay our experience of ourselves and of life. Unfortunately, these beliefs are generally unquestioned though they seem to underlie and support many psycho-spiritual and New Age systems. These unquestioned, infantile conclusions are antithetical to experiencing oneself as a human, and certainly experiencing (and I don't mean believing in) our quantum nature.

What "I" discovered was that infantile understandings not only prevented us from entering into our universal nature; but they called for a lifestyle which claimed a search for freedom but left us feeling frustrated and unfinished. In 1979, Nisargadetta Maharaj said to me, "Do you know yourself?" "I feel a lot of love and bliss," I replied, "And I can even see energy patterns." He yelled back, "I'm not interested if you're *satisfied* or *pacified* with your 'spiritual life'." And then he said with great disdain, "Do you know yourself?" I said, "No." He shouted, "Until you do, shut your mouth." In 1986, I began not only to talk, but also to present a way, through Nisargadetta Maharaj's enquiry, to undo what yogis refer to as the *knot of the heart*[1]. These *knots of the heart* represent pre-verbal, pre-conscious cognitions and conclusions which are the False Core.

First "I" saw that infantile trance states often underlie and support philosophies and religions which are based upon an infantile understanding of the world. For this reason, "I" will dare to dismantle New Age philosophy, religion and psychology itself with the hope of cultivating a much larger and broader understanding and experience of ourselves as the underlying unity.

---

[1] In *Trances People Live* the False Core (organizing principle) is also referred to by the Yogi's description, the *knot of the heart*.

A SPIRITUAL WARRIOR

It is not my intention to put down spirituality, psychology or New Age philosophy. But it is my intention to ask the reader to do what "I" myself have done, namely, to question concepts to see if they hold water. Or are they just more trances, conclusions or ideas taken from others which help to pacify, re-enforce and not experience present-time reality as present-time reality. Quantum Psychology looks deeply into everything to find the underlying premise. Once this premise is found and discarded, awareness can be further liberated and deepened. Certainly it is easy to appreciate that if a "system" holds a false premise (core) then those premises which follow must also carry further false conclusions. In short, if a system has a pivotal premise which is false, what follows it must also be false. Furthermore, a "system" based on a false premise and conclusion and whose solution is based on a False conclusion must also contain techniques which ultimately can only re-enforce their (the system's) own false premises. For this reason, if liberation of awareness is your purpose as a possible[2] aid in the discovery of **WHO YOU ARE**, then every belief must be held suspect and inspected, even your most "sacred cows."

Again recalling Nisargadetta Maharaj: "In order to let go of something, you must first know what it is." Or Ram Dass: "In order to get out of a jail, you must first know that you are in one."

With this in mind, I hope the reader does not take offense at what "I" propose but rather questions the belief systems that haven't been looked at, so that they can let them go and hence become even freer. It is not what we know about which is the problem—*it is the beliefs and conclusions we do not know about or are unwilling to question which create what we call pain which is a problem.* Our purpose is the liberation of awareness, not creating a life that "looks spiritual" is psychologically appropriate, virtuous, or healthy. Because of this, Quantum Psychology is often accused of being irreverent. But Nisargadetta Maharaj always showed *extreme irreverence to the False Core-False Self*—and simultaneously *extreme reverence*

---

[2]It should be noted that even now I am uncertain as to whether developing multi-dimensional awareness helps in the discovery of **WHO YOU ARE**. I imagine it could or might be an aid—but there are no guarantees.

to ESSENCE, the I AM, the VOID OF UNDIFFERENTIATED CONSCIOUSNESS and BEYOND.

This will not necessarily be a pleasant task and you might find yourself reacting with anger or fear. This is important because if you do experience anger or fear, it could mean you are banging into unquestioned identities and beliefs which threaten your False Core-False Self. As we continue to move deeper into the uncharted waters of our psyche, it is only through encountering concepts, identities and psychic structures that we can even begin to fathom the depth of our unquestioned premises which ultimately limit our freedom, limit our awareness and limit our experience of being fully human.

As with all true spiritual warriors, this should not be a concern. "I" am asking the reader no less than "I" have asked "myself": to question their concepts and notions. It is only through the questioning and the releasing of old concepts and structures that freedom[3] might be possible. Unfortunately, there are no shortcuts. Rather, it is through the deliberate understanding and letting-go of the *sacred cows* called concepts that spiritual and developmental trances can be released, and the true underlying unity of our quantum nature can emerge and become stabilized. To become a spiritual warrior, we must discover what precisely it is that we have believed or concluded—see it as just that, concepts—and ultimately "give it up." This is the enquiry which is *The Way of the Human.*

SPIRITUAL TRANCES AND DEVELOPMENTAL PSYCHOLOGY

In *Trances People Live*, the understanding of trance as a daily event, wherein we shrink our focus of attention and fuse with an idea, image, internal voice, body sensation, etc., was explored in great depth. In *The Dark Side of the Inner Child,* Chapter 14, "I" examined the process of spiritualization, in which we saw how infantile trances caused by developmental arrest or delays can cause and create spiritual philosophies.

---

[3]It should be noted too that even the concept of freedom needs to be questioned; and in Volume III even the concept of a concept is questioned.

The problem with spiritual philosophies is that all too often they deny our humanness. Asking us to label feelings as bad, or labeling sexuality or desires as bad, which further denies our humanness. To follow *The Way of the Human* requires only four principles: First, that you are human; second, understanding that your human nature is the simplest way to lead you to experiencing and stabilizing yourself in your Quantum Nature; third, that your human nature is not separate from your Quantum Nature; and fourth, that if you do not allow yourself to be human, then although you might have a taste of *Quantum Consciousness*, it will never become a permanent part of your awareness.

In Volume I, "I" mentioned that philosophies often stress behaviors which are not part of our animal, thinking or emotional nature, asking practitioners to hold high, lofty ideals which confuse the animal, the thinking and the emotional nature with **ESSENCE**. Unfortunately, these ideals, although beautiful in theory, can run contrary to basic human drives. Furthermore, to be human means to BE, without philosophy or ideal, just to BE.

In *The Way of the Human*, philosophies are viewed as being the result of developmental arrest which is caused by the shock of the Realization of Separation, thus creating the False Core-False Self (to be discussed later). To best explore this the reader is directed to the Special Section in Volume I, *Trances People Live Revisited*. When we begin to look at developmental theory in psychology as a lens and possible aid to point the way to determine what concepts we might unknowingly believe in, this might lead us to where we might be struck in "our" apparent psychological or spiritual dilemma. Unfortunately, the dilemma (i.e., our inability to stabilize awareness or find out **WHO WE ARE**), might be based on arrested development or trance.

With love
Your brother,
Stephen

PERSONALITY IS A MIS-TAKEN IDENTITY.

Nisargadatta Maharaj

CHAPTER I
THE FALSE CORE AND ITS BEGINNING

What is the False Core and how did it begin? The False Core is the one concept, the one conclusion, the one idea you hold which organizes your entire psychology. Freud said, "All traumas come in chains of earlier similar events." This simple and yet profound statement lies at the core of modern-day psychology. It says that the mind organizes events in *patterns*, a concept which Quantum Psychology finds extremely useful in describing the False Core. In Quantum Psychology we call this pattern the associational trance, namely, the tendency of the mind (nervous system) to organize and generalize events into similar categories. In this way, as a survival mechanism you make the past—which you now know how to handle—look like the present. For example, the mind—which at this level is a by-product of the nervous system—takes-on or fuses with our primary caretakers (i.e., mom and dad) and later generalizes them to our present-time relationships, attractions and repulsions, etc. Erickson called this tendency "associational networks"; and Korzybski would say that the nervous system (brain) abstracts (selects out) certain events and places them in different orders, the last in this order being the most general. My father was a man, my father was angry, my father was tall. Later, men who are tall get angry, all tall men are angry.

First, it can be said that the False Core is the organizer of all of the chains of earlier events.[1] So again, What is the False Core? It is the one belief, concept, idea or conclusion which is identified with which *holds*, not only organizes a single "associational chain" but is the organizer of all of your associations, thoughts, emotions, fantasies, etc.

For this reason it can be said, that the False Core *pulls your chain*.

Second, why does Quantum Psychology call it the False Core? Because the concept you hold about yourself and the world began to be solidified very early in your development. It was an assumption from the point of view of an infant and it was and is based on a *false conclusion*.

QUANTUM PSYCHOLOGY PRINCIPLE:

Any idea, belief, assumption, etc., which is born from an earlier false assumption, must also be false.

Simply stated, The False Core is false. Quantum Psychology uses the word for three reasons.

1. Because it is the core of everything in your psychology.
2. Because a core is what holds you up. If you are "held up" by a False Core, or a core which is not real but is a false illusory assumption, then you feel as though you have "no core" underneath you.
3. This False Core conclusion yields a False Self (which is the compensating part of the False Core) which tries to compensate, heal, hide, transform or overcome the False Core. The False Self—since it was created in reaction to a False conclusion (False Core)—is more false than the False Core itself

There is however "good news." Underneath your False Core lies your **ESSENTIAL CORE** (Volume III).

---

[1] This False Core "I" was discussed in my first Volume *Trances People Live: Healing Approaches to Quantum Psychology*, Chapter 20, "The Organizing Principle."

Chapter I

"We are not afraid of experiences, we are only afraid of what we have concluded that experiences mean."

Stephen H. Wolinsky

IN THE BEGINNING

How does the False Core occur? To appreciate this, we can use the lens or story of Psychoanalytic Developmental Psychology.[2] Their research and story says that newborn children believe that they and their mother are *one* and that between the age of five to twelve months, infants realize that they are separate from their mother[3].

It can be said (as a story) that the False Core *begins* to solidify at the time the infant realizes they are separate from the mother. The pain of the shock of the Realization of the Separation from mother is called the *narcissistic wound* or *narcissistic injury*. Quantum Psychology defines narcissism and its outcomes as "I am the center of the universe," "My mother and I are one," "I am the source of the universe," "I create it all," etc. Later it can become, "I must make mother my reflection" or "I must reflect her so she will reflect me, the world is my reflection, etc., etc., etc."

The False Core is the conclusion you draw about yourself and the world, based on the Realization of Separation. What the Realization of Separation means to you is, your False Core. For example, what you concluded about what "I am separate from my mother" *means*, might be "I am worthless" or the separation means "I am alone." I am separate *because* "I am inadequate." I am separate *because* "there must be something wrong with me." At the moment of this Realization of Separation, the False Core begins to solidify.

Quantum Psychology *theorizes* that the False Core is genetic-energetic and that it lies latent within the developing embryo, much like, if your grandfather had diabetes or heart disease, you too may have an genetic-energetic proclivity toward these health issues (to be discussed later).

The conclusion that you draw from that *shock* of the Realization of Separation from mom becomes the False Core. The shock is that suddenly you are separate from everyone else and from the rest

---

[2]Please note that Psychoanalytic Developmental Psychology is a lens, point of view, map or model of what is—it is not what is.

[3]This sometimes yields a pre-representational concept that they created their mother. Once this false conclusion is solidified, reasons for this creation by the infant, like lessons, Karma, etc., can easily follow.

of the universe. *The False Core is the reason or conclusion you draw about the separation from mom.* The False Self develops as a compensation in an attempt to compensate, hide, heal, overcome, resolve or transform, etc., your False Core. The False Self, for this reason, is an outgrowth of and an attempt to solve this False conclusion. In this way, the False Self is a part of the False Core. *This False Conclusion (False Core) and False Solution and compensation (False Self) seal your psycho-emotional and even spiritual fate.*

To explain further (in this story), your mother is your entire universe and you believe she and you are ONE. The fact that you are separate from her is a biological and psychological fact. When you realize you are separate, there is a shock, and then the rest of your life is dealing with the pain of separation by trying to overcome it, heal it, resist it, hide it, resolve it, transform it, spiritualize it, etc. The conclusion that I draw from the fact that I am separate is my False Core and I will continually try to overcome it in some way. For example, if I conclude from being separate that it is because "I am worthless," then I am going to try to overcome this feeling because it is associated and fused with the shock of realizing I am separate from the mom (later spiritualized as God the **ONE** a relationship, etc.). I will then resist the experience of worthlessness because it is associated and fused with the shock of the Realization of Separation. If "I am separate" means "I am inadequate"—and you believe it—then the compensation might be, "If I could overcome this by being smart, namely if I am smart, then I will be *merged* and ONE again and I won't have to feel inadequacy or defend against this anymore."[4] Bear in mind, *the False Core is a concept which you believe.* As mentioned earlier any *solution or concept based on a false premise must yield further false premises. For this reason, the False Self is falser than the False Core because it is a later concept (occuring after the experience) based on the earlier False concept or conclusion.*

It should be briefly noted that one of the failures of psychology is that if a client comes in with the problem of "I am worthless," it is assumed that this is truly the problem. Then all steps are taken to "work this out." However, Quantum Psychology maintains that "I am worthless" is a False Conclusion about an occurrence. And, that

---
[4]Please note all of this is far outside of one's Awareness.

you cannot heal a False Conclusion, because it is False. You must "see" the False Conclusion as a False Conclusion and discard it

QUANTUM PSYCHOLOGY PRINCIPLE:

Any treatment to try to heal or transform a False Conclusion is a treatment, therapy or spiritual practice which is organized by the False Self based on believing in the False Conclusion and, hence, can only yield a False Treatment, therapy or spiritual practice because the therapy or spiritual practice is being driven by believing in the earlier False Conclusions and premises.

QUANTUM PSYCHOLOGY PRINCIPLE:

The False Core is stronger than the False Self and therefore you will *always subjectively* prove that the False Core is true.

QUANTUM PSYCHOLOGY PRINCIPLE:

You can never overcome or heal your False Core, you can only be free of it by realizing it is not you.

BE MY MIRROR

How does this separation impact your life? For example, let's say you *"fall in love"* with somebody. When you first fall in love, you feel like you and the other person are *one*. This is the *honeymoon phase*. You meet them, you merge with them, you are one with them in this experience. When is the first time you have relationship problems? When you realize they are a separate individual from you. All of a sudden, one person says, "I want to go to the movies," and the other says, "I want Chinese food." It feels like a *shock*. Rage, anger, fear come up. Your False Core gets ignited and your chain of earlier events gets pulled, bringing up earlier similar shocks and associated with other Realizations of Separation, the most potent being the shock of the Realization of Separation from mom.

Chapter I

What happens is, whenever you have to deal with separation or differences, the narcissistic wound which began from the Realization of Separation gets re-opened and re-activated. To handle this the nervous system will attempt through an I-dentity to resolve the shock or chaos. This can be done through a myriad of psychological defenses like denial but even more deadening is the associational trance. Why is the associational trance more deadly? Because it makes things which are not the same appear to be the same, thus losing present-time relationships.

When this occurs the honeymoon phase is over.[5] It is a re-enactment of the shock of realizing that you and your mother are separate at the thinking, emotional, and biological level. Furthermore, you are no longer the center of the universe, you and your mother (later spiritualized as God) are no longer ONE—now you must try to get people to be *your* reflection and merge with you or be another's reflection to get them to merge with you in order to not feel the trauma of the shock of the Realization of Separation and the False Core. In the above example, this realization is generalized and later re-activated when you "get" that you and your lover are separate.

To further illustrate this re-activation of the shock of the Realization of Separation can occur if people do not mirror back to you your world, your image of the world or your image of yourself. This reactivates the narcissistic wound. The False Core can get detonated and go off like an explosion because the pain of the separation is fused (associated) with the False Core, which is fused (associated) with the shock of the Realization of Separation.

This resistance to the Realization of Separation can often get spiritualized. I once met a guru who said to her students, "The world is a reflection of you." Another said, "The physical form of the guru is a perfect mirror." This represents the guru's narcissism (as discussed in Volume I).

It is the resistance to the shock of the shock of the Realization of Separation which tries hard to see mom/the world as your reflection to avoid separation. The resistance to the Realization of

---

[5]Some theorists believe the honeymoon phase is an age-regression. Quantum Psychology feels that in the ideal honeymoon phase, you are aware of the separation and the **ONENESS** simultaneously (see Volume III).

Separation and the narcissistic wound causes your psychology, (i.e., thoughts, emotions, associations, etc.) to move to try to find a way to *overcome* your False Core. This (as will be discussed later) happens frequently, in both spiritual groups which give a spiritual practice which tries to overcome the False Core (i.e., variations on the theme of, "I am a sinner," "I have vices," or "I am bad," etc.) or in psychology practice where the False Core is re-enforced by the False Self using techniques like reframing, visualizing, reforming, reassociating, learning lessons, redeciding, accentuating the positive, converting vices to virtues, redemption to salvation, healing, transforming, etc. Simply put, when someone is not your mirror image, it is seen by the "I" which is a by-product of the nervous system as a threat to its survival and that is where the pain begins (i.e., fight, flight, escape, etc.), or the associational trance of *imagining this present situation or person is the same as that past situation or person.*

QUANTUM PSYCHOLOGY PRINCIPLE:
If you handle your False Core, then the narcissistic wound is handled. *The False Core is not you.*

QUANTUM PSYCHOLOGY PRINCIPLE:
In *theory* we are born with an energetic-genetic predisposition to a False Core. The Realization of Separation and its shock causes the narcissistic wound which begins to solidify the False Core.

THE NARCISSISTIC WISH AND PRAYER
When the pain of separation is so great the narcissistic wish and prayer both spiritualize and organize the chaos of the separation. Prayer is the narcissistic wish to merge or the infantile attempt to reach the omniscient mother (later spiritualized as God). Out of that wish comes your spirituality, your psychology and your trances.[6] For

---
[6] Please note this all being viewed through Quantum Psychology's lens of developmental psychology—it is a lens—a possibility to explore where you might through identification be stuck—it is not always or necessarily so.

example, "I will be able to merge if I do my mantra." "I will be able to merge if, *I do this spiritual practice.*" "I will be able to merge if, *I think the right thoughts"* (the power of positive thinking). "I will be able to merge if, *I achieve the right things, marry the right person,"* I will be able to merge if I (*fill in the blank*) etc. All are attempts to organize the chaos of the narcissistic injury.

SUBSTITUTION AND THE MERGER RESPONSE

Substituting basic wants and needs is what forms obsessive-compulsive behavior and compensating identities. It is an attempt to heal or overcome the narcissistic injury (the Realization of Separation).

Workaholics are an example of this substitution process. The question is, what is the underlying unmet need which is looking to be fulfilled? For example, somebody might have taught a workaholic that the way to get love (merger) is to achieve. Was the father overworked and an overachiever himself? He might have told or silently demonstrated that the way to get to heaven (i.e., find love, or merger) is to work hard or, if spiritualized, do Seva (Service) or Karma Yoga. That is why there is no real satisfaction in overwork. The obsessive-compulsive workaholic is not working because they love it, they are working to get an unmet need filled, which as we all know is a treadmill yielding no satisfaction.

Libido (as mentioned in Volume I) is the biological desire to merge. Everyone has a biological desire to merge. Substitution and pain occur because our parents and later society restrict our merger response.

QUANTUM PSYCHOLOGY PRINCIPLE:

Our merger response is prior to the False Core. It is biologically based.

QUANTUM PSYCHOLOGY PRINCIPLE:
The concept called "I am one" and later "I am separate" are both concepts because they both come after the non-verbal **I AM** (see Volume III).

QUANTUM PSYCHOLOGY PRINCIPLE:
Spirituality is born from the I am one *concept*.

QUANTUM PSYCHOLOGY PRINCIPLE:
Psychology is born to explain and handle the "I am separate" concept.

Narcissistic Merger:
The obsessive-compulsive drive to merge through substitution in order to overcome the Shock of the Realization of Separation.

Biological Merger Response:
The Natural Biological response to merge and separate.

QUANTUM PSYCHOLOGY PRINCIPLE:
The resistance to the natural biological merger-separation response is the Father and Mother of psychology and spirituality.

The restriction of our merger response is the origin of the substitution process. If you look at someone who is an economist, he might imagine he can merge through making money, which is a substitution for something else, i.e., an unmet need. Socially acceptable substitution can be unknowingly believing achievement equals merging. The implicit suggestion is that the way to overcome the narcissistic wound is by narcissistic merging. This is not the natural biological merger response which leaves someone feeling "complete" and at "peace"; but the *false* merger of the False Self which is a present time attempt to overcome a past time shock of the Realization of Separation and which requires action, activity and doing.

The natural merger response is effortless. Imagine your parents are not around for you. Then you put some toys together, and your parents say, "Oh, my God, this kid is going to be an engineer." For a second, you feel a sense of merger. "This feels great." The child decides if I build another thing or achieve something else, whether it be reading, good grades, etc., "I will merge." In this way, achievement then becomes the action, or the "doing" vehicle, to overcome the pain of the narcissistic injury (separation) so that you can re-emerge.

QUANTUM PSYCHOLOGY PRINCIPLE:
The promise of merger or union is implicit.

QUANTUM PSYCHOLOGY PRINCIPLE:
When you substitute a psychological want for a biological need in order to merge, you get no satisfaction.

> I can't get no satisfaction
> I can't get no satisfaction
> Though I try
> Though I try
> Though I try
> How I try
> I can't get no—
> No No No
> No satisfaction. . . .
>
> The Rolling Stones

SYMBOLS OF MERGER

QUANTUM PSYCHOLOGY PRINCIPLE:
Society offers you symbols of merger not merger itself.

A mother or father might "say" (overtly or covertly), "If you succeed, (i.e., have a big house), then implicitly you will merge and overcome the pain of separation." Those status symbols (houses, cars, etc.) become *symbols of merger*. In this case you have a symbolic substitution for the natural effortless merger response. The trance of substitution is defined as substituting a psychological want for a biological unmet need. *Unfortunately, the standards of merger are implicitly raised.* For example, I made $100 this week, my father said that's a good boy, you mowed 8 lawns and now you've got $100. But $200 is better (implicit message: *then you will merge*). So now I want to mow 20 lawns. Once you process your trance of substitution by realizing the unmet biological needs which yield the insatiable psychological wants, then you can see that you have not acknowledged the basic animal merger response.

QUANTUM PSYCHOLOGY PRINCIPLE:
You can only ultimately experience dissatisfaction and frustration through the false merger attempts of the False Self and the trance of substitution.

Why? Because no matter how many cars I have, I am still going to feel separate. And no matter how much sex I have, at some point—either just before I have an orgasm or just after—I am going to feel separate. The trance of substitution yields frustration.

When you are in the trance of substitution, you are substituting symbols of merger for the natural biological merger response. This creates pain because the symbol of merger is not the merger itself and will not overcome or heal the shock of the Realization of Separation.

Chapter I
EXERCISE

Step I: Notice a symbol you hold or held as of utmost importance in your life

Step II: Notice what biological need the symbol is trying to fulfill.

Parents implicitly shift a child's merger response by creating higher and higher expectations with the implicit message being, "If you clean your room, if you get better grades, if you go to college, you will get merger." It is oftentimes done implicitly, passed down from generation to generation. In addition, there are merger rules like "Sit up straight," look, act, be or do, and then you will merge.

The super-ego or "shoulds" come after the narcissistic injury. The super-ego represents the rules for how to merge, i.e., you *should* be good (if you are, you will merge.)

QUANTUM PSYCHOLOGY PRINCIPLE:
The obsessive-compulsive drive toward over-achievement is a substituted attempt to merge.

REICH'S CHARACTER ANALYSIS IN THE LIGHT OF THE NARCISSISTIC INJURY AND MERGER RESPONSE

Wilhelm Reich's character structure, for the most part, solidifies at a later age than the False Core. However, it does give an understanding which describes later developmental layers used to reinforce the earlier foundation of all psychology. Example:

Schzoid: The way to merge is through observation and infantile detachment.

Oral: Will continue to be oral because it believes that orality is the way to handle the narcissistic injury, i.e.,

"I can get my merger needs met orally or if I have no needs, I will merge."

Anal: Withholding or exploding is the way to frustrate another for no merger or to merge. "I will withhold and hurt myself or hurt another to get even for no merger and then somehow the other will "get it," see the errors of their way and want to merge."

Phallic-Hysteric: Genital or emotional is the way to merge, through sex.

Rigid: Achievement as the way to merge.

Psychopath: Power as the way to merge.

So what are your rules of merger? In other words, what implicit promise unknowingly drives you, i.e., that if you *do* this thing (i.e., get good grades, have lots of money, have a nice house, *(fill in the blank)* everything will be okay, which means you will be merged *again* somehow? What are they? Do you have to smile when you don't mean it? Do you have to play mom's or dad's game?

In 1975, I was a client with a therapist who used to do intense 2-3 hour Reichian and Gestalt therapy sessions. I remember getting to a point where I realized that in order for me to survive, I had to play my father's game. That was the realization and I began sobbing. But my tears were not like normal crying, it was more like a water faucet being turned on, which I have never experienced before or since. I had an emotional breakdown after that and couldn't do anything except stay in bed. It took me about a week before I could get back to therapy and to my life.

SHOCK POINTS

There are several *shock points*, the primary one being the realization that you are separate from mom which manifests as feeling separate from mom, the world, yourself and later, **GOD** or the

Chapter I

SELF. Another shock point is the realization that in some way you have to give yourself up in order to survive. But there are also many realizations of separation which come later and which are very dramatic, partly because they reactivate the original shock of the Realization of Separation.

Gurdjieff also used the term *shock points*. I believe he was really trying to provoke people's *shock points* or the shock of the Realization of Separation (their deepest Identities which organize around the False Core) and the narcissistic injury, the precursor of most pain.

Resistance to the shock of the Realization of Separation oftentimes yields a *pretend* to give myself up to mom or dad but, at the same time, to hold onto an age-regressed image of "me" so at least I can have a self. The problem with this is that these images are not you and the pretexts are not you. Soon you forget you are pretending to wear this mask (of surrender), and begin to believe you are the mask, image or persona because without them you would have no defense against the shock of the Realization of Separation. In short, survival solutions are: 1) I'll be just like mom or just like dad; 2) I will pretend to be like mom and dad (unfortunately, later you forget you were pretending);[7] and 3) I will not be like mom or dad *no matter what*. However, in the latter solution if I say I'm not going to be like mom or dad, I have to hold an image of them in my consciousness so I always know what I'm not going to be. This is why the mom/dad identity as models are so strong (to be discussed later).

There is another *shock point* of separation that occurs when a young girl or boy enters puberty and sexuality begin to emerge strongly, and all of a sudden, the father or mother creates distance. This *psychic shock point* can manifest itself later as a split between the heart and the genitals yielding, in Reichian terms the phallic or hysteric personality.

---

[7]Quantum Psychology uses a process of enquiry which is to ask, "Recall a forgotten pretend."

MAJOR SHOCK: THE LOSS OF ESSENCE

Newborn infants are essentially a blank slate. They exist prior to any conditioning.

During the Realization of Separation, something or someone must be blamed for the separation. **I AM** and **ESSENCE** which are now by the False Core False Self viewed as emptiness (as in a lack) are the only reference points and the only thing present at that time. For this reason, the spaciousness of **I AM** or **ESSENCE** is labeled as emptiness (as in a lack) yielding the belief *I have no core*. In this way **ESSENCE** *is blamed* and the conclusion drawn (False Core) is placed on the **I AM** and **ESSENCE**. In other words, the the **I AM** and **ESSENCE** which is now seen as emptiness (as in a lack) is fused with (no core) and hidden by the False Core. This *no-core* which is a conclusion placed on the mislabeled spaciousness **I AM** and **ESSENCE** is subsequently used to organize the chaos of the shock of the Realization of Separation and *"I have No-core."* This can yield a False Core (after all it is better than having nothing [no core] at all).[8]

For example, the shock yields "I am separate" *because* "I am worthless." But "I am worthless" because (I have no core) and this is because of this damned emptiness. If I could just get rid of this (heal it, transform it, hide it, spiritualize it, creates a new self), then I would not be worthless *anymore*.

This loss of **I AM** and **ESSENCE** makes us feel like something was lost and must be found for us to become whole or one again. This is the old story of *the necklace that was never lost*, it was around your neck the whole time. In other words, you are in a double-bind, unknowingly blaming and resisting the **I AM** and **ESSENCE**, because it is now mislabled as emptiness (as in a lack) and concluded that it means *"I have No Core,"* on the one hand, and seeking essential qualities like unconditional love (**ESSENCE**), on the other.

STRATEGIES TO OVERCOME THE INJURY

The shock of the Realization of Separation (5-12 months) is the strongest shock. Narcissism can be the outcome. Simply put, it

---

[8] It should be noted that many "therapies" interested in creating a *new self*, are born of resistance and re-enforce "I have no core or no self."

could look like this: 1) mom is separate, 2) that makes me crazy, 3) how am I going to get mom to merge with me and later be my reflection again, 4) I want mom to merge with me so I'm trying to get her to be a mirror reflection of me. Now since this cannot work, I can develop the strategy: 5) "Well mom seems to merge with dad." So if I can be like dad, then I can get mom to merge with me. In other words, at some point, you give up on mom and you try to merge with dad in the hopes that you can get mom to merge. For example, if an infant notices that when dad comes home mom gives him a hug, they assume mom merged with dad and, therefore, the strategy is, "I" will fuse with dad (use his strategies) so that I can get mom to merge with me.

It should be noted that many very recent "therapies" like NLP stress determining others' strategies and then *modeling* them in the hope of getting the "same" outcome. These "strategies" and the concept of "modeling" are rooted in developmental arrests and age-regressed states deeply developed and organized as a resistance to, or an attempt to, overcome or heal the narcissistic wound. This *strategizing* and *modeling* can lead to a deep grandiosity because the entire idea comes from an infant's state of mind which sees mom as magical, omniscient, omnipotent and able to do anything. Thus, if I adopt "her" *strategy* and *model* her (or dad), I can get and do what she gets and does. This can be seen in NLP specifically which attempts to model Mozart, Einstein and even Jesus.

Another strategy of the infant might occur out of an opposite situation—the *smother mother*. For example, you (as infant) are separate and actually it is okay. But mom does not want to experience her narcissistic injury when she had to realize her own separation. So mom is not going to let you be separate. Mom age regresses, makes you her mother, and now wants you to be her mirror reflection. You now have the little girl who becomes mom's mother (confidante). Mom tries in every way to get the little girl to be her mother and mirror her back. Einstein said, "Everything changes except for the way you think about it." In other words, oftentimes people do not see the external world and people rather they still see only their idea about the external world and people. In this way, they do not see the present, instead (unknowingly) they see only their relationship with mom or dad (the past) which they're trying to merge with and heal in the present.

VEHICLES OF MERGER

The rules of merger are basically how to play the game. If I play mom's game, then I merge. If I play dad's game, then I merge. If I am in a relationship with somebody and laugh at all their jokes, if I like Chinese food because they like Chinese food, I merge. In this way, there are certain unspoken rules of merger.

What are substituted vehicles of merger? It is the means or way of merging. It is the "socially" substituted and "appropriate" ways to merge. It can never yield long lasting satisfaction because you are trying to substitute the biological merging with (*fill in the blank*). For example, you are at a party and maybe you want to get close to somebody, put your arm around them, etc. That's part of the natural libido which was thwarted in some way during the socialization process (see Volume I, Chapter, The Biological Dimension). Instead, because of past associational learnings, you substitute the physical closeness and get into a discussion about politics or music.

Why not merge for the sake of merging as opposed to substituting which is merger with an associational network added to it. The social context has stamped out many of the natural biological drives. Society now determines what time we eat and how long I can hug someone, etc.

What's the vehicle you use to merge? One women commented in a workshop, "When my vehicle to merge doesn't match John's vehicle to merge, he doesn't like it."

You will enjoy and merge with people who have a similar vehicle to merge response. Society designates the form of "acceptable" merger vehicles which are appropriate for merger. The impulse to merge is biological. Our vehicles of merger get frozen. In other words, what are your particular vehicles of merger that you generally use? We generally have a small range. Note your substituted vehicles of merger. Is the the way you feel merged by *making money*? Is *doing* the way you merge? Is *status* a way to merge?

Chapter I
THE SPIRITUALIZED VEHICLE OF MERGER AND THE PATH AND RULES OF MERGER

You are separate at an external, thinking, biological, emotional level, and you've made a conclusion about what that means which is your False Core. The **PATH** to merger that your parents give you is an implicit message and promise that if you do certain things (what they tell you to do or imply you should do), then you will be able to re-emerge. So what are these rules of merger? What behaviors do you have to do? Is it, If I good grades, then I will be able to merge? Is it, If I have lots of money, a nice house, take long vacations, then I will merge? In short, If I do or have X, whatever X is, then I will merge. What are the rules of merger? Do I have to smile when I don't mean it? Must I play mom's or dad's game?

When these vehicles, ways and rules of merger become *spiritualized*, you get **THE PATH**. Spiritual rituals inknowingly reinforce mom and dad's path—you might get temporary relief from the pain of the narcissistic injury; but, often times, religions reinforce it. For example, shortly after a ritual if "I don't feel merged," well, maybe if I light four more candles, and bow down seventeen more times to the east in a particular way I will merge. Temporarily, during the ritual I get a sense of merger but, ultimately, it reinforces the injury. Why would it re-enforce the injury? First, the *spiritualized ritual* is done by the *False Self* in an attempt to overcome or heal the Narcissistic injury and shock, and, second, the False Self-False Core is holographic and one unit, so you can't have one without the other and one reinforces the other.

QUANTUM PSYCHOLOGY PRINCIPLE:
You cannot overcome the pain of the shock of the Realization of Separation—you have to go through it.

Now since the ritual does not and cannot work, instead of bowing down seventeen times, I escalate. Maybe if I do it eighteen times, the shock and pain will go away. Soon I imagine, "there must be something I am doing it wrong," that is why the pain and shock is still here. Please note, this is collapsing the levels. The pain is at the thinking, emotional, or biological level. "Spiritual" practice is a way to develop awareness of **ESSENCE** and beyond. Each "practice" is for a certain dimension. Although there can be overlap and they do interact it is important to know which practice will or won't affect which dimension. I once met a Buddhist and I told him about what I had been taught regarding processing reactions to death through Tibetan Buddhism. He was shocked. "I have been doing Buddhist practices for twenty years, I am working on doing 100,000 prostrations." But what do 100,000 prostrations have to do with anything. In short, mom and dad's way of merger gets spiritualized and **THE PATH** which people follow often times parallel and mirror the implicit/explicit rules set down in early childhood for merger.

LYING

The reason lying is so painful is because prior to the narcissistic injury, there is an implicit message and an *implicit promise* that take place. The promise is, "We were, or if you do, say or have the right thing (*fill in the blank*), we will always be merged," i.e., the child and the mother. That is not possible on a biological, thinking or emotional level because on those levels we are separate. Later there is a feeling that the mother (later spiritualized as God or the devil) has lied to me or I have fallen or been punished for original sin (the sin of separation from Mom (God)) which is bad and something I must have somehow done. Therefore, lying gets fused and associated with the narcissistic injury and the shock of the Realization of Separation. The rage or the outrage or the freak-out around this realization that we are separate feels like we have been lied to. *When someone lies, it can reactivate the entire shock of the injury.*

Demonstration

The purpose of this demonstration is to explore how the issue of narcissism impacts our relationship.

Chapter I
NARCISSISM AND RELATIONSHIPS

Joan is a 46-year old woman whose relationship just ended. She claims that *he* (Bob) is narcissistic. She is angry at him because he did not respond to her the way she wanted him to. "I" decided to focus on *her* narcissism.

Wolinsky: When you realized Bob was not a mirror image of you, what did you create?

Joan: I became a mirror image of him. I created this *persona* that would match him.

Therapeutic Note
Notice how in some forms of "therapy" there is the idea of matching[9] through verbal or posture responses to gain "rapport," thus creating a narcissistic illusion of merger with other.

Wolinsky: What did you assume, decide or believe that got you to create a persona that would match him?

Joan: If I matched him, maybe he would fall in love with me and I wouldn't be alone and I would have someone to merge with.

Wolinsky: When you realized Bob was not a mirror image of you, what did you *not* create?

Joan: Acceptance.

Wolinsky: What did you assume, decide or believe that got you *not* creating acceptance?

---

[9]*Psychotherapizing*: the process whereby a psychological procedure like matching, modeling or utilizing the transference, pacing, gaining rapport etc., is used to justify the developmental arrestment and delay of the *Psychotherapist*. It is used as a psychological philosophy, *unknowingly*, to resist and ultimately "act-out," the psychologists' developmental gap (this will be discussed as an age-regression) and destroys any possibility of either the therapist or the client getting out of the loop—in short, the therapist is age regressed and is justifying through their psychology their own developmental gap.

Joan: Well if he is not a mirror image of me, then I am alone.

Wolinsky: What's happening?

Joan: This is very powerful.

Wolinsky: When you realized Bob was not a mirror image of you, what did you resist?

Joan: I resisted being separate, I resisted being different. I resisted who he was. I resisted who I was.

Wolinsky: What did you assume, decide or believe that got you to resist that?

Joan: I resisted the False Core of being alone. (The False Core of alone will be discussed later.)

Wolinsky: What did you assume, decide or believe that got you resisting all of this and the False Core?

Joan: That I could die if he was not a mirror image of me.

Therapeutic Note

The shock of the Realization of Separation can feel like death or bring up deathlike issues.

Wolinsky: Where do you feel that belief in your body?

Joan: I feel it through here. (My face.)

Wolinsky: Notice the size and shape of the belief. Create an image of Bob over there (the other side of the room). Hold that belief like a lens, and look

Chapter I

	through that lens at the image of Bob. Now take it off and look at Bob. How does he look to you?
Joan:	Hard to see who he is when I do not look through the lens. He is pretty undefined. He is not very distinct.
Wolinsky:	And with the lens?
Joan:	He is defined and clear
Wolinsky:	Now put the lens on and have an image of your mother over there (the other side of the room).
Joan:	Okay.
Wolinsky:	Looking through the lens, how does she look to you?
Joan:	Well, I need her, like I can't live without her. She's very defined.
Wolinsky:	Now, take the belief lens off and put it aside. How does she look to you?
Joan:	Kind of undefined too, but not as undefined as Bob.

Therapeutic Note

Definitions and labels create boundaries and appear (give the illusion) to divide **THAT ONE SUBSTANCE**.

Wolinsky:	Okay, so from the point of view of looking at her now *without* that lens, she's more nebulous; with the lens she is defined.
Joan:	Yeah. You want me to do that with my papa?

Wolinsky: Why don't you do that now? Take the lens off.

Joan: He's less defined too. When the lens is on, I define him as "my hero."

Wolinsky: Good, now put the lens back on. If you define father as "my hero," what are you wishing?

Joan: I am wishing to be one with him. I am wishing to be merged with him. I am wishing to be inseparable. I am wishing to be fused with him.

Wolinsky: Now, if you define Bob like that, what fantasies do you make up?

Joan: That he will be there for me, that he will protect me, that he will take care of me, that he will meet my every need, that he will adore me, pamper me, baby me, nurture me, protect me.

Therapeutic Note

Notice the Trance-ference of DAD onto BOB.

Wolinsky: By having this lens on, what do you say to yourself?

Joan: "I can't make it on my own," "I'm helpless," "I'm vulnerable," "I'm defenseless," "I'm a little baby," "I'm insignificant," "I'm going to die," "I'll fall apart," "I'll disintegrate."

Wolinsky: Have an image of Bob next to Dad and say to Dad, filling in the blanks—please don't .

Joan: Dad, please don't *separate from me*.
Dad, please don't *ignore me*.

Chapter I

> Dad, please don't *neglect me*.
> Dad, please don't *avoid me*.
> Dad, please don't *distance from me*.

Wolinsky: Continue with, "I'll do anything, please don't (*fill in the blank*)."

Joan: I'll do anything, please don't *leave me*.
I'll do anything, please don't *reject me*.
I'll do anything, please don't *abandon me*.
I'll do anything, please don't *hurt me*.
I'll do anything, please don't *yell at me*.
I'll do anything, please don't *shake me*.
I'll do anything, please don't *scream at me*.
I'll do anything, please don't *hit me*.
I'll do anything, please don't *scowl at me*.
I'll do anything, please don't be *disgusted with me*.
I'll do anything, please don't *leave me*.
I'll do anything, please don't *avoid me*.
I'll do anything, please don't *resist me*.
I'll do anything, please don't *ignore me*.
I'll do anything, please don't *shame me*.
I'll do anything, please don't *ridicule me*.
I'll do anything, please don't *spank me*.
I'll do anything, please don't *walk away from me*.

Wolinsky: How does the core of your body feel now?

Joan: I feel pretty open.

Wolinsky: Okay, now look at Bob and look at Mom and Dad; now say to Bob, "I used you to avoid looking at my wound."

Joan: Yeah, Bob, I used you to avoid looking at my wound. I did. I used you to avoid experiencing the alone.

Wolinsky: How are you feeling right now?

Joan: Neutral. This is like the truth. Just saying what is.

Therapeutic Note

For Nisargadatta Maharaj, "you cannot let go of something until you know what it is" for psychology it is making the implicit explicit.

Wolinsky: How does it feel to hear yourself saying that?

Joan: Good. Yeah, I like overall telling the truth and just acknowledging it and getting it out there and this is what is. Taking responsibility for it.

Wolinsky: Now have mom and dad over there. I want you to start off with mom and *fill in the blanks* and say, "When I realized you were not my mirror image, I went ."

Joan: Mom when I realized you and I were not one, I went *to my father*.
Mom, when I realized that you and I were not one, I went *mad*.
Mom, when I realized you and I were not one, I went *crazy*.
Mom, when I realized you and I were not one, I went *sad*.
Mom, when I realized you and I were not one, I went *confused*.
When I realized you and I were not one, I went *away*.

Wolinsky: Now look at dad and say. "I try to be one with you and when that didn't happen, I went (*fill in the blank*)."

Chapter I

Joan: I tried to be one with you and when that didn't happen, I went *mad again*.
I went *sad*.
I went *scared*.
I went *lonely*.
I went *into alone*.

Wolinsky: "Dad, I tried to make you one with me by (*fill in the blank*).

Joan: Dad I tried to make you one with me by *being your little girl, by wanting you.*
I tried to make you one with me *by always smiling*.
I tried to make you one with me by always *doing nice things for you*.
I tried to make you one with me *by helping you*.
I tried to make you one with me by *always being there for you*.
I tried to make you one with me by *being strong like you*.
I tried to make you one with me by *disliking my mom the way you did*.
I tried to make you one with me by *walking like you*.
I tried to make you one with me by *talking like you*.
I tried to make you one with me by *being tough like you*.
I tried to make you one with me by *adoring you*.
I tried to make you one with me by *admiring you*.
I tried to make you one with me by *feeling sorry for you*.
I tried to make you one with me by *placating you*.
I tried to make you one with me by *organizing you*.

> I tried to make you one with me by *waiting on you hand and foot.*
> I tried to make you one with me by *anticipating your every move.*
> I tried to make you one with me by *hiding my anger and avoiding any conflict with you and by never disagreeing with you.*
> I tried to make you one with me by *imagining that you were this poor little boy who was neglected and abused and needed to be understood and I would be the one to do that for you.*
> I tried to make you one with me by *shaping myself to be what you wanted me to be.*

Wolinsky: Now say to Dad, in return, I expected you to (*fill in the blank*).

Joan: Dad, in return, I expected you to *adore me.*
Dad, in return, I expected you to *nurture me.*
Dad, in return, I expected you to *appreciate me.*
Dad, in return, I expected you to *love me.*
Dad, in return, I expected you to *take care of me.*
Dad, in return, I expected you to be *delighted in me.*
Dad, in return, I expected you to *look after me.*
Dad, in return, I expected you to *defend me.*
Dad, in return, I expected you to *protect me.*
Dad, in return, I expected you to *like me.*
Dad, in return, I expected you to *idolize me.*
Dad, in return, I expected you to *respect me.*
Dad, in return, I expected you to *cherish me.*
Dad, in return, I expected you to *honor me.*
Dad, in return, I expected you to *adore me.*

Wolinsky: I am going to add one, in return. I expected you to marry me.

Chapter I

Joan:	Yeah. In return, I expected you to marry me. Yeah that's embarrassing but true.
Wolinsky:	And when you didn't, I felt (*fill in the blank*).
Joan:	And when you didn't, *I felt betrayed. Yeah, I felt outraged.*
Wolinsky:	After all I have done for you, this is how you treated me.
Joan:	Yeah, really, after all I have done for you, this is how you have treated me. That whole thing I did with Dad, I did with Bob.
Wolinsky:	You did? (Jokingly)
Joan:	Yes, I did. Precisely. Exactly. To the tea.
Wolinsky:	I want you to look at your father one more time and make a statement, "I saw you as my last chance and I didn't want to blow it."
Joan:	Yes I did. "Dad, I saw you as my last chance and I didn't want to blow it."
Wolinsky:	Look at your mom. After I couldn't do this with you mom and it didn't work, I went and looked at dad because I couldn't do this with dad, that was my last chance and I didn't want to blow it again.
Joan:	"Since I couldn't do this with you mom, and dad was my last chance, I didn't want to blow it with him."
Wolinsky:	Because I knew I would never get another chance.

Joan: "I knew deep down in my core, I would never get another chance or let myself have another chance and I would be alone the rest of my life. And this was my last chance to not be alone." I was always wondering why I have done that. Put my focus on my father away from my mother. My mother got pregnant right away when I was a baby. And that she was busy with the other kids. They separated from me immediately. Mom was pregnant when I was six months old.

Wolinsky: She separated from you before you were ready—tell me something you decided, or assumed or concluded?

Therapeutic Note

In force theory, the early separation is a force (see Chapter XVII, Force Theory, in this volume) the child "feels must be resisted" (counter-force). It is not always a force coming at you. It can be a force or conviction taken away.

Joan: That I was replaceable.

Wolinsky: Since mom separated before you were ready, tell me something else you concluded.

Joan: That I wasn't who she wanted. That she didn't love me. That I was dispensable. That I was powerless and really alone.

Wolinsky: Where do you feel those beliefs in your body?

Joan: All over.

Wolinsky: Take the label off and have it as energy.

Joan: Okay.

Chapter I

Wolinsky: How are you doing now?

Joan: Larger. More spacious, more whole.

Wolinsky: Okay. How are you going to get even with mom?

Joan: By going to dad. Resisting her.

Wolinsky: And stealing her husband?

Joan: Yeah, stealing her husband. Yeah, get him to love me more than he loves you (Mom).

Wolinsky: Okay, how you doing now?

Joan: (Laughing) That's pretty crazy and wild. I feel great.

REVIEW

"I will mirror you [the man] and I will become your reflection, so that you will merge with and mirror me," is the theme. If you want tea, I will take tea. If you want something, I want it, too. Though she said in therapy, "This guy was so narcissistic," I was more interested in *her* narcissism. If we trance-late, her behavior looks like this: "I will mirror you (Bob) *so that* you will mirror me." When he (Bob) does not mirror her back, she gets pissed off. Unknowingly, outside of her awareness, she unknowingly has,[10] "I can't believe he doesn't give me what I want, after all, I mirrored him and gave him everything he wanted. Now, he's separate, he's showing individual-

---

[10] The unconscious is formed to repress unwanted experiences. This is why I disagree so strongly with the Erickson or Jungian idea that "you can always depend on your unconscious mind to give you what you need." Quantum Psychology suggests that you can always depend on the unconscious to set you up to repeat your patterns. If you think about it, if you could always depend on your unconscious mind to give your what you need, integrate, etc., then why don't you always pick the right relationships, jobs, situations. The idea that the unconscious gives you what you need is a re-frame and a defense against dealing with the pattern directly. Instead of dealing with it directly, we add a sugar-coated layer of, "I must have needed it," to justify the pain.

ity and I'm pissed off." Simply stated; in a narcissistic relationship, *"I'll mirror you so you will mirror me."*

TRANCES AND MERGER: THE NARCISSISTIC WISH OR RE-ENACTMENT NOTES

SELF TALK AND THE NARCISSISTIC WOUND

Internal dialogue, or self talk, is a trance inducer and sometimes either re-enacts the Narcissistic Wound or tries to heal it. Look at what is being said in your self talk. It often does one of two things: 1) It tries to heal the Narcissistic Wound (i.e., talk about possible merger); or 2) it says things like "I fucked up," "I'll never (future) get what I want" (i.e., merger). Discover the *process* (merger) of your self talk rather than the content (he said/she said). In other words, if I am using a trance of future orientation in time with myself, is it forecasting merger? Some therapies have you imagining a time in the future when you will have (<u>*fill in the blank*</u>) or even use thought affirmations. These are unconscious attempts by the False Self to attain or reach merger through a substitution, For example "I am worthy to get money," which unconsciously equals merger and survival. Furthermore, this is an attempt of the False Self to have or get something as a way to overcome the False Core and the Narcissistic Wound or the present time pain of the Shock of Separation (to be discussed later). It is an attempt to unknowingly overcome the narcissistic injury by doing or having, as if substituting something for the shock could overcome the shock. As will be discussed throughout, this appears therapeutic but only re-enforces the False Core.

For example, oftentimes, as a therapist a client would present a problem like, I'm married and my spouse is not giving me what I want but she/he is in therapy and I know in the future everything is going to be really great in our relationship. This is using the trance of futurizing to handle the pain of *separation* by forecasting future merger. *The question is, Are your trances anticipating merger or failure to merge and separation?* Is forecasting possible merger *a wish* or is it imagining separation and thus duplicating the pattern of the

Chapter I

separation of the Narcissistic Injury?[11] This will reveal the distractions which defend against the narcissistic wound and which organize around the False Core.

QUANTUM PSYCHOLOGY PRINCIPLE:
What happens in the world (content) is secondary to how you interpret what happens.

In other words, you interpret events through your False Core and so if a person broke up with me do I interpret it as meaning "I am inadequate," "There must be something wrong with me," "I don't exist," etc.

THE SPIRITUALIZED REALIZATION OF SEPARATION: TRANCE-PERSONAL TRANCE-FERENCE

The myth of the fall of man/woman from God (i.e., the separation from God) because he/she did something wrong (original *sin*) and the concept of redemption (re-merging) are spiritualized and, ultimately, defensive metaphors to explain this natural Realization of Separation—and the desire or wish to merge again.

Spiritualization occurs when parents are trance-ferred onto God. This is a trance-personal trance-ference. Teachers, therapists, and gurus oftentimes implicitly or explicitly make promises of merger (like mom/dad, etc.) to hook clients and disciples. These promises are so powerful because they are pre-representational, infantile, and remain unquestioned to the student, client, or disciple. They are unquestioned for two reasons: 1) they are so early they are simply accepted as the way things are; and 2) just as it is bad to question Mom/Dad, teachers, therapists, or gurus etc., so it is not okay to question or not approve of a teacher's, therapist's or guru's actions. For example, I knew a disciple whose guru slept with underage girls. When I said this was inappropriate, she responded by saying, "You have a lot of stuff with this teacher." "You mean if I think it's inap-

---

[11]See *Trances People Live* or The *Dark Side Of The Inner Child*.

propriate for a 74-year old man to sleep with 13-15 year old girls who are his students, it's my stuff and my problem, not his?" She didn't answer.

NEW AGE SPIRITUALITY

New Age spirituality can also be based on narcissism and grandiosity. Imagine an infant in a crib. The infant's parents are around the crib. The infant smiles, the parents smile, the infant cries, the parents frown—the result, "I create, cause or I am the source of my parents' behavior." Furthermore, there is the narcissistic distortion and implication that "the external can be controlled by my internal" or "I am the source, and the world organizes around *me* and *my* beliefs." This is collapsing the external and thinking dimensions. Rather than understanding "my beliefs" might help create or justify my own subjective internal experience and act as a lens for me to view the world and how I imagine the world views me, but I am not my beliefs or lenses.[12]

SPIRITUALIZATION—
MORE TRANCE-PERSONAL TRANCE-FERENCE

Transpersonal trance-ference was discussed in depth in Chapter 14 of *The Dark Side of the Inner Child*. Transpersonal transference can also manifest when the external mom/dad are spiritualized into Gods and Goddesses which are internalized and subsequently spiritualized as a way to resist the Realization of Separation. Later, there might be an attraction to Buddhism where people take refuge or place a Buddha inside of themselves, i.e., "I take refuge in the Buddha" "as if" the Buddha were inside of you. In this way because of developmental issues there is a mis-understanding. *There is nothing inside of you that is not also outside of you and taking refuge in the Buddha is taking refuge in* **THAT ONE SUBSTANCE** (see Vol-

---

[12]It should be noted that imagining your beliefs create reality is neurologically inaccurate. In other words, experiences happen. Then, neurologically later, a belief is formed to justify why this or that occurred. In fact, the belief comes after the fact and only justifies why things occur, to organize the chaos by giving a reason thus giving a person an illusion of control; i.e., since this belief caused that to occur, if I change my belief, then I can change what occurred.

Chapter I

ume III). If you have an image inside of you it is oftentimes the way you are resisting feeling or knowing something. Imagining internal Buddhas, Gods or Goddesses can re-enforce, and add a layer to and over, the shock of the Realization of Separation and the accompanying False Core. At best they are very primitive, developmental structures which act as Band-Aids over the narcissistic wound. At worst, they add another layer to the earlier pain. Most people have this tendency to resist the shock of the Realization of Separation and to internalize images of Mom/Dad and later spiritualized images of Goddesses and Gods. In this way, they unconsciously imagine they can overcome the shock.

To explain further, prior to the shock of the Realization of Separation, an infant holds an archaic representation of merging. When the shock of the Realization of Separation is spiritualized and mom and dad are made into Gods and goddesses, you imagine you are spiritual by trying to merge with the Guru mother or father (Ammaji and Babaji in Hindi). Actually you are holding onto an infantile pre-representational experience of merger which you are now unknowingly *acting out*. Holding onto this experience is the way you resist the shock of separation which naturally occurred. Another way to look at it is; There is an archaic pre-representation of *you* merged with mom. You resist the shock of the Realization of Separation and hold onto the archaic representation of merger. Next you get involved with people, i.e., it could be lovers and/or teachers, therapists, gurus (mothers or fathers). The idea is to merge with them in some way "imagining that it is spiritual." Actually, *you are "acting out" the resistance to the biological and psychological shock of the Realization of Separation and calling it spiritual.*

The more you resist that separation, the higher probability that you are going to go into the *trance of spiritualization*. There is no fault or blame with that. *Spiritualization* is more or less intense depending on the context (what occurred or did not occur) during the the shock of the Realization of Separation and the pain of that separation. For example, a schizoid personality had to deal with the Realization of Separation in-utero (to be discussed later). That is why the schzoid False Core—*I Do Not Exist*—is what you might see most often in "spiritual" groups.

In relationships, you can do it with a lover. "I just want to merge with you." "I just want to disappear with you." To differentiate this acting out of the Realization of Separation with Spiritual Realization we can talk of Rumi. God was his lover. He was not "acting out" of an archaic pre-representation, but in recognition of the underlying unity. This was not a reactment of the separation anxiety at the biological level. If he were "acting out," then the experience of ecstasy which he describes would be extremely fleeting.

QUANTUM PSYCHOLOGY PRINCIPLE:
In India, the old saying is, "The bigger the Guru, the bigger the Maya." Maya is illusion.

With this understanding, sometimes (particularly recently), teachers, gurus, therapists, etc., create the illusion that they can lead you to merger (relabeled as health which will lead to merger)—this is an illusion. In other words when I am healthy then I can merge. The bigger the guru, the bigger the Maya seems to occur because the illusion of merger with another (acting-out the Realization of Separation) has to be told again and again and then exaggerated as the number of people increase.

QUANTUM PSYCHOLOGY PRINCIPLE:
The Guru/Disciple, teacher/student, therapist/client model is, at best, an attempt to confront and heal the Realization of Separation and go beyond. At its worst, it leaves people in dependent, regresive states and in even greater pain.

In the Guru/Disciple model, the promise is of merger. Those teachings are: "Focus on me" (the guru), "fuse with me" (the guru), "fixate on me" (the guru), "become one with me" (the guru) or "be like me" (the guru) and "you will merge and be free." Simply put, if you fuse with me, then you will merge with everything and no longer will you have to experience the pain of the shock of the Realization of Separation. Instead you will realize the unity of all. This in no way

Chapter I

implies that everything is not made of one substance. However, THAT ONE SUBSTANCE cannot be realized through an infantile **I**-dentity. The painful side of the identity is the shock from the Realization of Separation. To overcome that shock a compensation is the seeker's belief that if I become the guru (I will be omnipotent, omniscient, like magical Mommy). If I become enlightened (the mommy side) then I won't have to experience the pain of the wound anymore (the infant side of the polarity). Simply put, one side of the polarity is, "I am wounded and separate." The other side is, "I am **ONE**, omnipotent, omniscient, etc."

If you have a Guru who sits on a throne and asks you to mirror them (play their game), then you have age-regressed adults who mirror (play the Guru's game) by acting like them. This reinforces the narcissistic wound because the wish (fantasy) is "If I just do this, or be or reflect dad/mom (Guru), then I will merge again." It is also interesting that Gurus are often referred to as "the perfect mirror." It should be noted, that Gurus, teachers and therapists who make these implicit demands are exhibiting an integrated age-regression. This means he/she has found a milieu to act-out, in a "socially acceptable way," their age-regressed counter-trance-ference.

SPIRITUALIZING NARCISSISM: THE PATH OR WAY

Notice if you have a fantasy of merging with a parent and if you are doing the **WAY** of the parent or the **PARENT PATH**, then ask yourself, Do I *somehow* spiritualize this with Gurus and God? In other words, if you follow their path, parents promise Nirvana—a house, 2.3 children. Later the spiritualized trance promises Nirvana, bliss, etc., if you follow the path of the Guru. The promise of enlightenment (where all pain ceases and you merge with God/Mom) can be similar and mirror the implicit promise of merger made by parents which later becomes spiritualized.

THE PATH OF THE PARENT

In the early 70s, when I was in group therapy, traditionally you had about ten people in the group, 50 percent of whom were women about the age of 25-35 who had done everything "right."

They married the right man who had a master's degree or more, they lived in a nice house and had 2.3 children. Then, one day they suddenly realized, "Oh my God, I feel so shitty? I don't get it. I did everything I was supposed to do but I didn't get the outcome" (Parental Nirvana). In other words, they had followed the rules of merger and were still suffering. Often pop-psychology *hooks* people's (substituted) symbols of merger (i.e., money, attractive partners, etc.) by saying, "If you do or get (*fill in the blank*), this workshop process, you can have it all" (merger). This is a re-enactment of the **PARENTAL PATH**. Money and a more attractive partner are symbols and compensation for the merger which did not happen in the past. They are attempts to make-up for unmet biological needs. These status symbols which represent merger and can never yield satisfaction

QUANTUM PSYCHOLOGY PRINCIPLE:
To try to substitute a biological merger with a psychological symbol (i.e., money, status, how people view you, houses, etc.) for a natural biological need is collapsing the levels (i.e., trying to substitute the External or thinking levels for the biological).

SPIRITUAL PATH
Spiritual paths are oftentimes, "What do *I* have *to do* to re-merge?" That is why ritualized spiritual practice often only reinforces the narcissistic injury. It is an attempt to merge. Your emotional pain oftentimes leads to the Narcissistic Wound and the False Core so you can use the pain to "break through to the other side."[13] *There is no way out of the pain other than through it*. But with awareness, you *can* navigate through.

To sum this up, trying to overcome the Realization of Separation through substitution or through the external collapses the levels. It is trying to get the external to handle the pain of the internal condition. The Narcissistic Injury and the shock of the Realization of

---

[13]See Introduction to *Hearts on Fire: the Tao of Meditation* (Wolinsky, 1995).

Chapter I

Separation are now internal psychic memories and you cannot successfully use external spiritual rituals to overcome them.

In this way, oftentimes because of trance-ference people do not see the external world; they only see their relationship with mom or dad. They continually try to use others in present time to heal injuries that occurred in the past.

QUANTUM PSYCHOLOGY PRINCIPLE:

During the shock of the Realization of Separation, you lose the **I AM** and **ESSENCE** and identify with the body shock. At the moment of shock, the False Core begins to solidify and, simultaneously, the loss of awareness of the **I AM** and **ESSENCE** occurs.

At the time of shock, all levels collapse. That's why, when people are around Gurus, teachers or therapists they try to merge with them, an unfortunate practice which sometimes gets encouraged. Merge with me, think of me, fixate on me, put all your attention on me. This is collapsing the **VOID** with the Guru's thinking, emotional, biological and external dimension. This is why, when you are around students or disciples of a Guru, oftentimes they look, act and pick up the Guru's mannerisms, just as they once did with mom and dad, assuming that it is somehow spiritual or will lead to merger.

NARCISSISM AND SPIRITUALITY

The False Core will always reinforce itself unless you are fortunate enough to have somebody willing to *confront* you at all costs. A "good" Guru should do this, but most cannot because they have their own narcissistic wounds and they *need you to be fused with them.*

If a Guru or therapist has not handled their own shock of the Realization of Separation and Narcissistic Wound, when a disciple, student, or client wants to leave, they label them as bad and might tell them, "You haven't surrendered, you're resisting, you have a lot of worldly karma to work out." In Quantum Psychology, a teacher should naturally mirror back **ESSENCE, I AM** and the **VOID**, etc., and, simultaneously, allow for and acknowledge external, thinking, feeling, and biological separation, and simultaneously frustrate and confront peoples self-deception. This should occur naturally, without internal considering, shoulds, or as a standard. For the teacher rather it is just what is.

Spirituality becomes distorted because you can never experience the underlying unity through a False Core-False Self. If you are, it's going to be temporary and age-regressed. This is why people around Gurus act like children (*age regressed*) expecting a "Magical omniscient, omnipotent Mommy" or "Daddy" to take care of them and give them the blessing of merger. In a more mundane way, let's imagine falling in love with someone. If you cannot or do not see their separatness and individuality, then you are age-regressed. Though your age regression has brought about a *temporary* illusion of unity, when you realize they are separate or different, this infantile merger disappears and uncomfortable emotions emerge. This is the shock of the Realization of Separation. The Narcissistic Injury and Realization of Separation must be explored and dissolved, otherwise another's individuality can never be appreciated.

NARCISSISM AND THE NEW AGE

In many ways, the New Age exhibits narcissistic tendencies. I remember one New Age Guru saying "The universe gives me lessons. It *reflects* back to me what I need." This is collapsing the levels of the external with the "uncooked seeds" of the narcissisim of the thinking and emotional dimensions. Whatever seeds you haven't cooked will pull you out of the awareness of the **ESSENCE, I AM**, etc., and later belief systems will be created to explain why I "lost" or came out of the underlying unity. Soon techniques *guaranteed* to bring you back to **ESSENCE, I AM** are created but they do not understand that *there is no guarantee*, no "I" can go back to the **VOID of UNDIFFERENTIATED CONSCIOUSNESS**—the **VOID of UNDIFFERENTIATED CONSCIOUSNESS** takes you back (see Volume III). It is not in the **VOID** or **ESSENCE** where the belief system is created, it occurs when separate "I" awareness begins to re-appear out of the **VOID**. This is when you come back to your psychology. The False Core-False Self then comes up with reasons for losing **ESSENCE,** or the awareness of the **VOID** in the same way as an infant comes up with reasons, why me and mom are not one anymore.

INFANTILE TRANCE-FERENCE AND THE GURU'S COUNTER TRANCE-FERENCE

Guru counter trance-ference:
"If you do what I want or say, you will merge and get what you want."

Guru mirror counter-trance-ference:
"If you mirror me, act like me, etc., you get good things or what you want. If you don't mirror (surrender) you are resisting are bad and won't get what you want."

—Disciple trance-ference: Looks for mirror (re-enactment of narcissistic injury)
—Guru counter-trance-ference: Throws infantile rage at disciple (child) when they don't mirror back what the Guru wants.
—Double trance-ference: Guru age-regresses and becomes a child wanting the disciple to mirror them (play their game). The child wants a mom/Guru mirror or world which reflects and is one with them.

I have oftentimes seen this with psychotherapists who imagine that they are not age-regressed because they are "appropriate." They are good girls or boys saying and doing the "right" thing. This is dangerous for them as therapists because it is an *integrated age-regression,* i.e., thinking and "manipulating" to look and act appropriate by observing other's minimal cues, then modifying their behavior to match another's behavior. This is an attempt to use their developmental delays in an acceptable context.

Recently I met a therapist from Colorado whose integrated age-regression was so deep, he was on automatic diagnoses. He thought he was taking training (psychologizing) to be a better therapist. However, he was doing it unknowingly to re-enforce his integrated age-regression. Furthermore, he would diagnose his clients and tell them his diagnosis. In this way, he would appear even more magical and omniscient. But neither he nor his clients—some of whom had been seeing him for ten years—were aware of his counter-trance-ference integrated age regression, or his desire to be and appear as "magical Intuitive Mom."

What made Nisargadetta Maharaj so unique was he would bust the trance-ference and age regression. That is what he did with me. I was age regressing and had trance-ference issues and did not know it. I thought the game and the way to merge was, whatever the Guru wanted me to do, I would do. When I met him, he said, "Are you willing to stay eight days and absorb the teachings?" "Whatever Maharaj wants me to do, I'll do," I said expecting him to say, "Oh, that's great. Let's keep this guy around here bringing me tea for the next five years." But he said, "I don't do that. If you want to stay, stay. If you don't, don't. I don't play the Guru game." I went into shock. The shock was my Narcissistic Injury. I didn't even know I had a spiritual game and was age-regressed. He was able to kick my ass (False Core) and reinforce the UNDERLYING UNITY simultaneously. He was enabling me to grow more aware of **ESSENCE-I AM** and the **VOID** while he was clobbering the identities which I didn't even know I had. I had thought I was spiritual because I was doing the India spiritual game. In reality, it was a *spiritualized* re-enactment of the merger transference (see Special Section *Trances People Live Revisited*, Volume I).

NARCISSISM AND THE SMOTHER MOTHER

Parents frequently use their children to not have to feel their own False Core and narcissistic injury. A parent can transfer their own parents onto their children, turning their children into their parents. They then treat their children as their parents and merge with them so as not to feel their own separation. This is how you get the *smother mother*. Every time the child tries to be independent, this

Chapter I

type of mother will not let them because if she does, she will have to deal with *her* own Realization of Separation and narcissistic injury. A smother mother can come up with all kinds of stories to justify her behavior but these are only stories to defend against her pain.

In relationships people feel separation. They have a hard time because it can be a reenactment of their narcissistic attachment to their mother. What they conclude during separation (the False Core) will demonstrate how they react. Do you get depressed? Do you go into rage? What do you do to not deal with that separation? What you do and how separation is handled lead to the False Core. For example, in private practice as a therapist, oftentimes a woman (a wife) would go into therapy and begin to become independent and separate. When that occurred, the husband would go into his Realization of Separation (narcissistic injury) and hate the therapist even though they might never have met. This is a classic re-enactment of the separation from mom.

FREUD REVISITED

Freud got himself into trouble by misusing terms (i.e., like sex) and not placing them in the context of a child's years (age). For example, the tickle that a one-year old feels between their legs is not the same as sex to a person who is forty. In this way, by not specifically being clear about the context of it, Freud decided the tickle was the same as, and meant the same to, the child as it did to someone who was forty years old

Second, Freud suggested or implied that the libido was purely sexual. Quantum Psychology suggests that this natural biological merging response is not only and always sexual but is the biological desire to merge with at one or more of the levels.

Because of societal taboos and repression of sexuality, etc., this merger response oftentimes gets confused and associated with sexuality itself. Once this occurs, sex is substituted for this "natural" biological merger response.

In "modern" psychology, there is a lot of confusion between sex and intimacy (i.e., the desire for and the ability to have a merger response). For example, recently a "popular theorist suggested (the great American "come on") that as you get older, you can have more and better sex. Quantum Psychology would suggest as you get older your *quality* of sex and your ability to merge and be intimate might possibly increase. But the *quantity* will probably decrease.

Children wish and desire merger with mom. Not sex. Why? Because at that age, a child does not know sex, what sex means or even possess the biological ability to have sex; however, they biologically know merger. Being unable to "merge" the child looks outside and receives "shock" number two. The father seems to be merged with mom. This shock leaves the child—who is trying to organize the chaos of the shock of the Realization of Separation—with the idea, If I fuse with dad I can become one with mom. Shock number two deepens the False Core and can bring up many emotions and wishes—whatever that means to a child (probably fusion or merger).

Thus the child might compete with dad or other siblings for the desired object (mom) but it is a mistake to call *all* of this sexual.

THE FUSION ASSOCIATION TRANCE— THE TRANCE OF THE NARCISSISTIC IMAGE

For years, I wondered why people made everything the same as everything else. For example, I know many people who claim Quantum Psychology is the same as everything else from Psychodrama to NLP or you (Stephen) are just like (*fill in the blank*). At first, this confused me until I understood that they have to make everybody and everything the same to avoid the pain of separation of their narcissistic injury. According to Korzybski, this tendency of over-generalization is a survival mechanism contained within the nervous system.

QUANTUM PSYCHOLOGY PRINCIPLE:

> The fusion trance occurs when the mind makes everything the same as everything else to avoid the chaos of a separate differentiated object (person, place, situation or thing).

For example, if everything and everyone is *separate and differentiated, then the shock of separation is reactivated.* Therefore, it

Chapter I

is less painful to see everything as the same. This resistance to seeing differences and differentiation can get mirrored in the body with the lack of differentiation in a person's ability to move.

QUANTUM PSYCHOLOGY PRINCIPLE:
You can be "in love" with someone as long as you see differentiation. If you cannot see or know difference at an external thinking, emotional or biological level, you are not in love, you are in an age regression.

THE TRANCE OF THE NARCISSISTIC IMAGE

Love is blind when you imagine that you are in love but you see no differences. This is age-regression and a mirror trance-ference (see Special Section, Volume I).

A lot of people in relationships get angry with the other person because "I mirrored them, and after all that I have done for them, they will not mirror me back. What and who do they think they are, individual and separate?" Many of the problems people experience in a relationship are because the other person is acting differently than the image they have of them. If you have an image of someone. and they do something different from that image, people don't like it. But when I hold an image of someone, I lose present-time reality, and I lose who the other person is at the external level in present time.

The trance of the narcissistic image happens when you shrink your focus of attention, lose external reality and treat the external person like the image you have of them, which is the one you fell in love with. Then you try to get them to mirror back to you the image you have of them. When they do not, you cannot understand this and become frustrated or upset. What compounds the problem is that you then feel you have something to discuss with them, something that needs to be handled or worked out or that they need to change so that your image of them is met and you do not feel pain. As though they might have something to do with the fact that they are separate and different from your image. In some strange way, unless this is addressed directly, you still expect them to reflect back to you your image. This is *the trance of the Narcissistic Image*.

QUANTUM PSYCHOLOGY PRINCIPLE:
The conclusion you draw about someone or yourself for not living up to your images or you not living up to their images, is the same conclusion you drew at 5-12 months.

TRANCES OF THE NARCISSISTIC WOUND

We psychically return to the narcissistic wound and the shock of the Realization of Separation unknowingly again and again, and resist it, simultaneously, by using the trance of *Age Regression*. Freud's repetition compulsion is the overwhelming tendency to repeat patterns. Here we keep going back (age-regressing) to the trauma for two reasons: 1) to try to *right the wrong* of the event because we are resisting the perceived chaos of the earlier situation; and 2) because there is a deep-seated belief that if we go back in thought, (i.e., obsessive thinking) or by re-enacting emotions in therapy. This can be likened to repetition compulsion), that somehow we can now do it right. It all is contingent on resisting what occurred. Some therapies actually have you "go back" and *right the wrong*, i.e., do it the way you would have like it to be, which reinforces the age-regressed wish.

TRANCES AS RE-ENFORCERS OF THE NARCISSISTIC INJURY

Trances can re-enforce, distort and buffer the experience of present-time reality (see *Trances People Live*).

Below are some common trances held in the context of the shock of the Realization of Separation and the Narcissistic Wound. In this way, we can begin to understand the depth of trances, the why and how of the motivation and how they are connected to the basic survival mechanism within the nervous system.[14]

---

[14]In Volume I, The Biological Dimension, emphasis was placed on how biological deprivation, threatening survival activates the survival mechanism within the biology.

Future Orientation:
When you go into future fantasies: 1) are they positive fantasies to avoid the pain of the Realization of Separation; 2) if they are negative fantasies, do they repeat through resistance the repetition-compulsion to *right the wrong*?

Post Hypnotic Suggestions:
When you talk to yourself through internal dialogue do you create merger or justify and, hence, reinforce lack of merger?

Dissociation:
Caused by a resistance to the separation realization.

Over Identification:
Caused by smother mother's age regression, i.e., not allowing the infant to be separate. This helps the smother mother to resist her separation.

Time Distortion:
Freezes the separation realization and slows it down by time distortion as a resistance to the separation realization. This paradoxically highlights the realization, slowing it down and making it more painful.

Confusion:
Confusion resists "what is" about separation realization and acknowledging differences.

Amnesia:
The denial of the knowledge of the separation realization.

Sensory Distortion:
Numbing out the pain of separation realization.

Negative Hallucination:
Not wanting to see, hear or feel separation or merger. The latter is because merger is fused with the Realization of Separation.

Positive Hallucination:
Seeing, feeling, hearing or imagining separation where there is none or seeing, feeling, hearing or imagining merger where there is none.

Hypnotic Dreaming:
Dreaming, wishing, praying of/for merger or separation.

The Observer Trance:
Strong dissociation to avoid the pain of separation. Followed by Sour Grapes—"I didn't want to merge anyway," Sweet Lemons—"Separation is what I really wanted," or *Spiritualization*; observation separation is more spiritual (the Yogi Trance).

THE OBSERVER (TRANCE)— THE TRANCE OF THE YOGI[15]

From a Quantum Psychology point of view, the observer occurs by dissociation from experience. This means there is something which happens which cannot be processed and the observer is formed diminishing the awareness of the biological dimension to defend and dissociate from this unwanted experience. This is discussed in greater detail in the False Core-"I Don't Exist" which has the strongest proclivity toward being an over-observer. For now, let us say that the "I don't exist" False Core is formed earlier than all of the other False Cores (in theory, since it cannot be proved as a fact). It is called schizoid by both Wilhelm Reich and Alexander Lowen, a student and client of Reich's. Lowen says that the schizoid is formed in-utero and is contained within the bony structure of the body. It is caused by the rejection by the mother. Quantum Psychology adds to this that the dissociated observer is formed at this point to defend against this rejection.

This is why the False Core of "I don't exist" is so dead set against "feelings" and the "body" and oftentimes leads to the attraction people have to join certain spiritual groups or systems which claim that: 1) the body is somehow bad, feelings are bad, something to be purified, etc. For example, I once knew a psychiatrist from

---
[15]See Volume I.

Chapter I

Nevada who was anorexic-bulimic. He spiritualized this by claiming he was purifying himself through fasting and laxatives, rather than resisting feelings, sexuality and the body. 2) "I am not the body." This should not be confused with the "I AM NOT THE BODY," which is a realization at the **NOT-I-I**, whereby the body is "experienced" as the same substance as everything else (see Volume III).

The observer trance takes place at the thinking-emotional-biological level and has a tendency to "think" experiences rather than to "have" experiences.

For example, I know an Erickson trainer who is schizoid with a False Core of "I don't exist." When I would ask him how he was, he would say, "Carl Jung says, _____," or "Joseph Campbell says, _____." He once said to me, "Whenever I feel upset, I immediately create all the resources I can." "What's wrong with being upset?" I asked. I also knew a guru wannabe from Texas who wrote a "spiritual" book. When I read it, "I" laughed because he was a student of Swami Muktananda and "his" experiences were the same as Muktananda's in *Play of Consciousness*. He did not know the difference between having experiences and having fantasies and thoughts about experiences he had just read. (We will return to this in the False Core, "I Do Not Exist.")

THE DISSOCIATED VOID (THE TRANCE OF THE YOGI)

One of the most complex trances is the *dissociated void* (see Volume III). The *dissociated void* trance occurs when a child is traumatized. To resist, the child places a "blankness" or "fuzzy emptiness" on top of the trauma.

Since no trance defense is foolproof, the trauma continues to leak through the "*dissociated void.*" The *dissociated void* looks like the **VOID**, but it is a trance.

How do you know if it is the **VOID** or the *dissociated void*? In the former, there is no issue or "I" and nothing "leaks through." In the latter, there is a sense that "something is not right," soon "something" (meditation, fasting, silence, ritual, mantra etc.) *must be* done more and more by the False Self, like a repetition compulsion (unsuccessfully trying for union with mom now spiritualized as GOD).

CONCLUSION

In this chapter, I have attempted to explore a possible theory as to where the False Core arose from, i.e., the shock of the Realization of Separation. But whether this theory is true is actually secondary. What is primary is that when the biological drive to merge—or any other biological drive, i.e., eating, sleeping, going to the bathroom, having sex, etc.—is thwarted in some way, the the body's basic primordial survival mechanism kicks in. Specifically, with the separation-merger, it appears to mirror not only psychological processes but also spiritual seekers looking for merger.

This is in no way a condemnation of psychology or religion. But as we embark on the 21st century, it offers us an opportunity to re-evaluate and explore what is motivating us, and what either succeeds or fails, in psycho-spiritual practice. Because what seems clear is that we lose present time when we live in past time with unprocessed infantile concepts, conclusions and trances.

CHAPTER II
THE FALSE CORE: AN OVERVIEW

What is the False Core and how does it work? The False Core is a false premise. It can be considered energetic-genetic and it might begin to solidify between 5 to 12 months of age.

The False Core is the basic primary concept you have about yourself which like glue holds your entire psychology together. This false premise-conclusion-assumption-idea organizes your every thought, emotion, action and reaction, etc. In 1986, in the original version of *Trances People Live*, I called this the *Organizing Principle*.

It is crucial to explore the False Core because if you have and believe in a false premise, then all of the premises which follow must also be false. It's as if you were building a house (your life) on sand and trying to compensate for this false foundation by using bricks (psychology, spirituality, etc.) to keep it together.

According to Freud, "All traumas come in chains of earlier, similar events." This statement suggests that everything in our psychology is organized and associated in what appears to itself to be similar. It can be said that for survival reasons, the nervous system has a generalizing function. For example, "all men are" (*fill in the blank*), "all women are" (*fill in the blank*), "all relationships are" (*fill in the blank*), "all races are" (*fill in the blank*), etc.

But the nervous system (to be discussed later) makes things appear as similar and generalizes, making them the same when they are not. This survival mechanism is built within the nervous system. This leads us to several Quantum Psychology Principles:

The Way of the Human • The False Core and the False Self

QUANTUM PSYCHOLOGY PRINCIPLE:
Any action, thought, idea, emotion or reaction, etc., based on a false earlier premise, can only lead to future false premises.

QUANTUM PSYCHOLOGY PRINCIPLE:
The nervous system organizes the mind in chains of associations or patterns to perpetuate and continue its own survival.

QUANTUM PSYCHOLOGY PRINCIPLE:
It is the organization in chains of "earlier" events—which the nervous system makes appear similar but which may are not—which creates the concept of a pattern.

QUANTUM PSYCHOLOGY PRINCIPLE:
It is intrinsic for the body-mind's nervous system, as a survival mechanism, to organize patterns where there are none.

QUANTUN PSYCHOLOGY PRINCIPLE:
The False Core or false premise can only lead to false conclusions and false assumptions about experiences because they are not based in present time reality.

QUANTUM PSYCHOLOGY PRINCIPLE:
If you are fused with the False Core, then most of your entire internal subjective experience will prove or be set up to prove your False Core as true.

Chapter II

To paraphrase Oscar Ichazo, the founder of Arica:

> Since all my experiences are based on past accumulated [premises], I cannot perceive the present except in terms of the past.

The False Core is the primary concept we have about ourselves, and it organizes our entire psychology. It organizes our view of the world, our view of our self, and how we imagine the world views us. This False Core has an emotional component also, but (as was discussed Chapter I, The False Core and its Beginnings), it was solidified, abstracted and concluded during the shock of the Realization of Separation between the ages of 5 and 12 months.

To begin this section, I want to talk very generally about what the False Core is and what it means to you. As we continue, we will go into greater detail.

THE ENNEAGRAM

The Enneagram of personality is a description of the movements of the mind. Quantum Psychology suggests that all of the personality types and their movements are driven by the False Core. The personality types exist as a means of resisting the False Core and are driven by the False Core. In short, no False Core, no personality types.

The Enneagram is a tool to look at behavioral, emotional and thinking patterns. However, Quantum Psychology is not interested in spending a lot of time in typing personalities and discussing a myriad of ways the False Self compensation acts. For Quantum Psychology, these pursuits are distractions and take you away from direct confrontation with the False Core. What is primary is realizing that you are not your False Core Driver or False Self which attempts to hide and compensate for the False Core Driver. Thus, Quantum Psychology looks at 1) What is driving the personality type; 2) what compensates for your personality type; and 3) dismantling the False Core which drives the personality type. It should be noted that Quantum Psychology considers over-discussion of types and over-typing part of the associational or fusion trance and, hence, a defense and a resistance to experiencing not only the False Core and the chaos which occurs because of the shock of the Realization of Separation but also present time reality.

QUANTUM PSYCHOLOGY PRINCIPLE:
Anything you think you are, you are not.

QUANTUM PSYCHOLOGY PRINCIPLE:
Whatever you know about, is not you.

QUANTUM PSYCHOLOGY PRINCIPLE:
Anything you think or imagine yourself to be, you are beyond.

The question then is, "Why talk about and observe who you are not?" Because how you fixate your attention is important, not only so that you can know what it is, but so that you can take it apart and go beyond it. In Quantum Psychology, we find out what our organizing structure is around which we organize every thought, every behavior, every emotion and every everything. This was called the *Organizing Principle* and a yogi would understand that to cut this associational trance and False Core cuts the *knot of the heart*, thus liberating the Awareness of **ESSENCE**.

The Enneagram can be used to understand *your personality type—not in typing another*—but then it should be discarded. Quantum Psychology explores the driver of the personality and its dissolution and then once beyond this, Quantum Psychology too should be discarded. There is a Zen saying that "the finger that points at the moon is not the moon." Describing the Enneagram of personality is a beginning step, it is not the goal. It can re-enforce the personality when we declare I am a (*fill in the blank*). This misunderstanding arises when we worship the finger (the Enneagram description) rather than the moon (**ESSENCE-I AM**, etc). In the words of Alfred Korzybski, "The idea or description is not the thing it is referring to."

Discovering the False Core which drives your psychology might help you find out who you are or as Nisargadatta Maharaj said, *"You find out who you are by first finding out who you are not."* The Enneagram of personality types describes the way you fixate your attention to overcome the False Core. Quantum Psychology suggests that the driver of your personality, the False Core, must be discovered and dismantled in order to realize **WHO YOU ARE**. Do not worship the Enneagram or Quantum Psychology. There is no Enneagram path or Quantum path. People think they are their fixation. Quantum Psychology studies the False Core to demonstrate how the personality is/was organized in order to understand who we think we are, but we must come to realize we are not.

The purpose of Quantum Psychology is to find out *WHO YOU ARE*. The purpose of discovering your False Core is to find out *who you conclude or imagine you are but are not. In this way, you can find out who you are by finding out who you are not.* As mentioned earlier, Nisargadetta Maharaj used to say: 1) "In order to find out WHO YOU ARE, you must first find out WHO YOU ARE NOT." And 2) "Anything you think you are, you are not."

In this process, we will work on developing eight dimensions of awareness (the **VOID** and the **NAMELESS ABSOLUTE**, which are beyond dimensions and beyond Quantum Psychology, erase the eight dimensions (see Volume III)).

A man (woman) is three things: 1) What others think he (she) is; 2) what he (she) thinks he (she) is; and 3) what he (she) really is.

          Anonymous

Chapter II

QUANTUM PSYCHOLOGY PRINCIPLE:
Don't confuse what you think he (she) is, he thinks he (she) is, with what he (she) really is; and

Don't confuse what others think you are, you think you are, with what you really are.

THE RISE AND FALL OF THE ENNEAGRAM

The Enneagram of personality types is thousands of years old. It was *very, very briefly mentioned by Gurdjieff and it was brought forth and greatly elaborated on by Oscar Ichazo*. Quantum Psychology regards Ichazo as the Father of the modern-day Enneagram. From a Quantum Psychology perspective, the purpose of the Enneagram is to find out who you think or imagine you are so that you can let go of this imagining, go beyond it and discover **WHO YOU ARE**.

The Enneagram became available to a popular audience in the mid-1980s with descriptions and on-going discussions about personality types. People used to say "I am a two," or "I can tell you're a three," etc.

The never-ending descriptions reminded me of what the late Alfred Korzybski, the father of General Semantics once said, "You can always say more about what you said."

Two major problems arose from this. Only Ichazo in the late 1960s and 1970s and Quantum Psychology in the 1980's came up with ways out of or beyond the False Core. Most other groups at best offered observation, which is a beginning step but which ultimately cannot work. Why? Because the observer is part of that which is observed. And at its worst, attempts were made to fix, transform, convert, make healthier or change the False Core which cannot work. Why? Because *all attempts to change the False Core are done by the False Self. The False Self seductively makes one believe it can change or heal the False Core. However, the False Core-False Self is part of and is holographically one piece or unit.*

Furthermore, the descriptive and frozen tendency to use the I AM a (*fill in your number*) left us with a linguistic problem; it implied that WHO YOU ARE is who you were, are, and will forever be.

ABUSES OF THE ENNEAGRAM

Many people do not understand the depth and level of the False Core-False Self, and how difficut it is to "go beyond it."

To illustrate I met a psychologist from Vermont who was a disciple of the Enneagram. At the same time, he was in an integrated age-regression which meant that during his childhood, in order for him to survive his "crazy" mom, he had to psychically leave "his" world, figure out and "appreciate" mom's inner state and give her the appropriate response. In cognitive therapy, this cognitive distortion is called "mind-reading." This complicated system led him to Enneagram teachers who gave courses in typing, understanding and appreciating the nine personality types. In this way "you" could "deal" with people appropriately. Unfortunately for him he was re-enacting his childhood with his mom and has now found a "spiritual" and "psychological" way to continue this age regression.

In his case, as a child he could mind-read Mom and give the "appropriate" response—now he could use the Enneagram as a justification to help him mind-read and re-enforce his integrated age-regression. In other words, his attempts at typing were a more sophisticated and socially acceptable way to re-enforce his age-regression.

THE RISE AND FALL OF DIAGNOSIS CHARACTERIZATION AND THE ENNEAGRAM

Sometime around the second year of a child's life, a discovery is made that along with colors, sizes, shapes and other attributes "things" have names: Observers report that during this period, the child begins to point at objects, asking the equivalent of "What's that?" and when told the name of the object appears to be quite satisfied that he has learned something about the object. (Irving, Language and Habits in Human Affairs, p. 155).

That is when the child realizes that a name can be given to anything, and that name may become the most pronounced character of the things. (Koffka, The Growth of the Mind.)

He may go on to assume unconsciously 1) that when he learns a name for anything, he somehow gets to know the important characteristics of whatever the name is associated with; 2) that the uniqueness of the "thing" is revealed by the name; and (3) that if a "thing" has no name, it is either nonexistent or inconceivable. (Irving, p. 155)

"Piaget's findings lead him to suggest that the child does not regard the name "as being inscribed on the thing" but that the word and thing are fused. The name is in the object, not a label attached to it, but an invisible quality of the object. (Piaget, The Child's Conception of the World.)

People act as if words (diagnosis and typing) and what they represent have some inherent inseparable and perfectly natural connection. This leads to the implicit assumption that the "thing" is just what the word(s) says it is, and conversely that words may even take the place of "things." In brief, the effect of this childish identification of verbal and nonverbal levels often leads people to substitute the language for the facts of life. (Ibid 156-157)

In other words, there is an illusion that the label placed on a person (their diagnosis on type) is who they are and that the words themselves have some kind of intrinsic meaning.

The fall of the Enneagram of personality further lies in its descriptions of each type. The Enneagram does not describe what drives the psychology (i.e., the False Core) but rather it describes its defense and compensation to the False Core. For example, most fixation descriptions like "the perfectionist" describe the compensation that you use to try to *"over-come"* your False Core, not what is driving the mechanism.

THE I AM (FILL IN THE BLANK) DISASTER

The I am disaster is based on the disaster of the words "I am." David Bourlands, creator of E-prime, had a brilliant understanding which eliminates the verb "to be"—be, am, are, is was, were—from the English language.

> When we say that something *is* something, we dishonestly, although probably unconsciously, suggest that what we *think* must really exist. This hides the fact that we use *to be* to make judgments and to put people in pigeonholes by describing them with a single adjective or noun." (International Society for General Semantics, (1991). *To be or not to be*. San Francisco, p. 10).
>
> Alfred Korzybski said it this way, "Every time we use 'is' we lie" (Ibid, p. 14).
>
> and
>
> A map belongs to a 'level of abstraction' different from that of the territory it represents.

In other words, the description belongs to the *thinking level only*, it does not describe who this person is. When you use descriptions at the conceptual level, you do 3 disastrous things:

Chapter II

1. You rob the person of many human qualities which the description does not include;
2. You remove the person further from who they are (in your mind) at a Quantum Level;

and

3. You rob yourself of a multi-dimensional interaction with another human-being because you begin interacting with them through your labels and descriptions which are primarily at a thinking level, thus missing who they are in present time.

> "A 'map' does not contain all of the structural characteristics of the territory it represents" (p. 30).

In short, "The map is not the territory." "The idea is not the thing it is referring to." The *person who is diagnosed, typed or labeled is not who they are but rather a miniscule description of who someone thinks or imagines they are at a thinking level only.*

A beginning step is to use the Korzybski idea of dating and indexing. In other words, rather than saying, "I am a (*fill in the blank*)," you can say, "In 1950, I used the concept of (*fill in the blank*) to describe myself." This gives a *date* (1950 in space-time) when the concept of being a (*fill in the blank*) was created which then organized your psychology

EIGHT DIMENSIONS OF AWARENESS

THE TWO DIMENSIONS OF NON-AWARENESS

Along with most forms of psychotherapy, Quantum Psychology does focus on the thinking and emotional dimensions. This is because if your awareness is continually fixated in one particular dimension, then it will be "gobbled up" by how and where it is chronically and habitually fixated. If all your attention is on the thinking dimension, you won't be able to connect with your physical body and you will lose the external world. If you are so fixated on your relationship with your mother or father (a past representation which exists in memory), then you are also going to lose your awareness of the present-time external world, your physical body and your present-time relationships.

Awareness is literally eaten up by fixations. If all of your awareness is focused on the concept that you are worthless and you fuse with that concept, then you will lose access to the other dimensions of awareness. This chronic concept "you" use as a lens to view "you" and the world, imagining this is how the world sees you and its compensation, is what Quantum Psychology calls your *False Core-False Self,* respectively.

QUANTUM PSYCHOLOGY PRINCIPLE:
In order to stabilize your awareness at the level of **ESSENCE** or **I AM**, how you fixate your awareness (i.e., on your False Core-False Self) must be processed, digested and liberated.

The problem is you cannot go beyond your False Core through your False Core or False Self. The False Core is invisible to you and it is invisible because you think it is you. It is characterological because you are unaware of it. For example, "I *always* feel inadequate," "I *always* feel worthless" or "I *always* feel like I am alone." These False Cores are never questioned. They are transparent, like looking through a glass of which you are unaware. In the same way, you see the world and "yourself" through this lens which you think is you, hence, the lens goes unnoticed.

GOING BEYOND THE FALSE CORE
You cannot go beyond the False Core until you are willing and able to acknowledge it, be it and to sit in it totally. When you are able to sit in your False Core, then you can move beyond and sit in your **ESSENCE** or **I AM**. But you have to know what pulls your chain of associations. You can sit in meditation and have an experience of **ESSENCE, I AM** or **NOT-I-I**; but when you open your eyes and go out into the external world, your psychology and the world

Chapter II

are going to smack you in the face. For this reason, you have to be willing (*and you cannot will yourself*) to be in your False Core.

If your False Core is "I have an inability to do," and you try to get out of it in some way, it is always through resistance. Therefore, you have to be willing[1] to *be in your False Core without the intention of trying to get rid of it or change it in some way*. Whenever you try to get rid of the False Core or even transform or heal it, then you have resistance to it. Unfortunately, much of spirituality and psychotherapy falls into the False Self's trap by trying to maneuver you away from this False Core even going so far as to ask you to keep the positive and leave the negative. This False Self compensation is seductive and can only add to the defense resistance and re-enactment because it unknowingly re-enforces the False Core.

BLOWING THE FALSE CORE

When the False Core blows, it does not mean your psychology ends. All of your psychology, as well as your thoughts and feelings, will continue, but they will not have any ground to land on because there is no "you" associated with it any longer. In other words "your" psychology becomes very distant. For example, even if there is an enormous amount going on in your external thinking, emotional and biological dimensions, when your False Core is blown—even in the midst of fear, pain, or anger—you will still feel unaffected and beyond it. In other words, *there is no subjective experience of reactivation because there is no you, because there is no False Core-False Self.*

GOING BEYOND THE FALSE CORE

If you can observe your fixation *without any intention of getting rid of it*, then it is not an avoidance. The question you have to ask yourself is, "Am I observing the False Core to get rid of it, change it, heal it, transform it, etc., or am I observing the False Core to just observe it and notice what happens."

---

[1] Like the first de-labeling exercise, energy and emotions in Chapter 4 of *Quantum Consciousness*,

YOU ALWAYS TAKE YOUR FALSE CORE WITH YOU.

If, as Freud maintained, "All traumas come in chains of earlier similar events," then according to Quantum Psychology, "The False Core is what pulls the chain of associations."

By observing someone's outward behavior, it is easy to imagine that you understand their False Self or False Core. But to determine what this actually is, namely, what pulls the whole psychological chain, is a difficult endeavor. You have to understand the *motivation* behind their behavior, what drives their psychology. Let's say someone is studying in school. Are they studying because they "feel worthless and inadequate?" Or because "they don't want to be alone?" Do they "want to have power," or do they believe "there is something wrong with them?" This is why it is hard to know what the False Core is.

The False Core is what drives you and pulls your chain. It will always reinforce itself; all False Cores reinforce themselves. But when the False Core is gone beyond, your psychology continues running but there is *no you* there to grab it.

The False Core is resisted, thus the False Self emerges—but also the False Core is *desired* like a black hole with a powerful force that pulls everything in it—*the False Core interprets everything through itself.* It *desires* its survival and, hence, its repetition—and as a by-product of the nervous system, it does not want to die.

A student once asked Nisargadatta Maharaj, "If you are enlightened, why do you still smoke?" "Years ago I left my human nature to take care of itself," he replied, "What do I care?"

FROM WHAT OR WHERE DOES THE FALSE CORE ARISE?

The False Core might be seen as an energetic-genetic pattern that you "take on" or absorb along with its corresponding compensations from your parents and lineage. You fuse with this genetic-energetic pattern, which is pre-verbal and pre-representational. It is a combination of the energies of mom and her lineage, and dad and his lineage, mixing together that give you a lineage. Since your particular False Core might have a genetic aspect, there can be a predisposition to it.

Under high "stress" situations such as conception, things are more intense and there is a greater tendency "to take things on." Some infants will "take on" the False Core of their mothers as the only way to have a relationship with them. An important question that needs to be asked is, *"Prior* to taking on this False Core, was there any chaos that occurred?" Because during a chaotic situation, there is a greater tendency (within the nervous system) to reach for something to organize and handle it.

QUANTUM PSYCHOLOGY PRINCIPLE:

The False Core is false. It is not you. It is not your real core. It is a concept which acts as a lens which you are operating and living out of.

THE FALSE CORE IS A FILTER

If someone has a False Core, everything is filtered through it—how they see themselves, how they see the world, and their projection of how they *imagine* the world sees them. This is true for every False Core.

How your attention gets fixated is a decision, conclusion or a reason given which becomes solidified during a shock trauma. If I conclude I am "unworthy," for example, this means that most of my attention actually becomes fixated on "unworthy." *This is the fixation.*

If I try to compensate in some way, this compensation is known as the False Self. Even while I am trying to compensate by proving I am worthy, or to heal or transform the unworthiness, I am unknowingly fixating on "unworthy" and trying to compensate for it. If, for example, you are fixated on "unworthy" and try to overcompensate for it by "over-giving," still deep down you believe you have no value (worthless). All of your attention is unconsciously and unknowingly on worthless-no value. If my False Core is "I am incomplete," I am going to try to become complete or whole in an attempt to overcompensate and overcome "I am incomplete." But deep down I still "feel incomplete."

Without realizing it, most of my awareness is on *incomplete* and recreating incomplete through stories or spiritual trances to justify or explain it in some way. Quantum Psychology focuses on *how* to liberate your awareness from your fixation of the False Core-False Self in hopes of increasing the possibility of really "getting" **WHO YOU ARE**.

THE FALSE CORE IS MIS-INFORMATION

The False Core is your subjective source of misinformation as well as the vehicle by which you interpret all of the other dimensions. Belief in (mis)information is a major problem. It begins with I am (*fill in the blank*). But *it* unknowingly recreates itself again and again, always winning, always reinforcing I am (*fill in the blank*). "All False Cores are misinformation." For example, if you had a belief that "all men were bad and all women good," that is misinformation. Nisargadetta Maharaj used to say that "Anything you know about yourself came from outside of you, therefore discard it."

When I was growing up, my parents used to say, "You aren't any good in math," even though I had scored 95 out of 100 points on New York State Regents Exam. As a result of this, when I was a junior in high school, I scored 76 out of 100. That misinformation I "took on" is how I began to see myself and "act out" of. You always set yourself up in situations that will reinforce the False Core. This is why, when you are put in situations that do not reinforce the False Core, it can feel quite threatening. You feel you're on red alert as if your survival were being endangered (since you think you are the False Core and the False Self which are hooked into and part of the nervous system with its survival, fight-flight mechanism).

When the identities get in touch with what's beneath them (**ESSENCE, I AM**, etc.), there is an incredible panic. The identities know they are going to die. The False Core False Self is merely a veneer and lens that covers **ESSENCE** (also known as the **ESSEN-**

TIAL CORE (see Volume III)). And when the False Core blows, you (**THE BIG YOU**) permeate the space, going through all of the body. Awareness of the **VOID** begins to expand and the person is filled and feels large and peacefully aware (See Volume III).

QUANTUM PSYCHOLOGY PRINCIPLE:
You cannot get out of the False Core from inside of it. You cannot think your way out of the False Core from inside it.

THE FALSE CORE DOMINATES: PROCESS VS. CONTENT

To illustrate the False Core process, a woman came to see me for therapy with a whole agenda. She had been involved with a man for years but the relationship had ended. I did not work on this problem because her False Core was, "I don't exist," and I already knew what kind of relationships she would form and who she would be attracted to. In this example, since she did not exist (subjectively) she was attracted to a man who fit her M.O. (modus operandi); hence, she tried to fit into his world. If she did not handle the False Core in therapy, this pattern would be simply repeated with the next person she was involved with. So there was no reason to spend hours talking about this man. Obviously, I knew what he was like. It should be noted that for the most part we are attracted to a person who re-enforces our False Core-False Self.

Quantum Psychology focuses on process, not content. Many therapies claim that they are process oriented, but I have seen very few that truly are. Process oriented means I'm not *that* interested in what happened and I'm not *that* interested in your memories. What I am interested in is *the process you continually go through again and again*, whether it be with mom, dad, siblings, lovers, husbands or wives. It always comes back to the same questions—how do you organize and what is the False Core that you organize *everything* around?

BEYOND THE FALSE CORE: NO INTERNAL CONSIDERING

When you go beyond the False Core, to use Gurdjieff's terminology, there is no more *internal considering*. Everything is just as it is and you are no longer mentally distracted. In other words, you are doing what you are doing when you are doing it. There is no standard like, "This is good for me or bad for me." There is no judgment, evaluation or significance placed on events or situations. You realize that *the story you (the nervous system) make up comes after the experience has already taken place.*[2]

The fact is, the story about what occurred or what a situation means arises after they are over. This story becomes a reason and a justification for the re-enactment of the False Core-False Self. If you are in present time, you are not in any fixation or dimension and there is nothing going on. For example, you go out swimming. Are you experiencing the water or making assumptions about it? *Each False Core determines and interprets what an experience means.* Does it mean "I'm inadequate?" Does it mean "I'm not enough?" Does it mean "there's something wrong with me?" *When you are in present time, there isn't any fixation. You are just there in what is happening.*

---

[2] It should be noted again that Nisargadatta Maharaj called the "I" the *body-mind*. This was because the "I" and MIND are by-products of the body and nervous system and are inseparable from them. In other words—No-Body or nervous system, no-"I."

Chapter II

"There is no good or bad. There is just our thoughts about it."

Hamlet
William Shakespeare

The Way of the Human • The False Core and the False Self

FINDING MEANING

Trying to find meaning is a function of the nervous system's False Core-False Self complex. Things mean what you (nervous system) make-up or imagine they mean. Actually, situations, events, emotions, thoughts, etc., have no intrinisic meaning. They mean what you fantasize they mean. If I say you look nice today, you could say that means (*fill in the blank*). You (nervous system) interpret everything through the False Core.

You even interpret what your illnesses mean to you through the False Core. What does the sickness mean to you? Does it mean, "Oh God, I really screwed up? Am I furious because I really feel powerless?"

When you interpret and assign a meaning to what is going on, you make inferences about what is—which is not what is. Let's say John tells you "you look nice today." If you assign a meaning to what John said (i.e., he said that to make me feel good because I really look bad) that is a meaning and a story that you assigned to the situation. But you don't take any responsibility or acknowledge that you (nervous system) made it up. Now you feel upset because you know deep down that John said you look nice because he really thinks you look terrible. So now you have a whole story going on and you are in pain because of it (conclusion made after the experience). You decide I'm not going to go out with him anymore or have anything to do with him because he always tries to make me think I look good when he knows I don't. What you have done is make up a whole story and placed a meaning on "you look good" which creates actions and reactions, etc. Unfortunately, you take no responsibility for the meanings "you" placed on his remark.[3]

FINDING YOUR FALSE CORE

Your emotional states are pointers to the False Core. In the introduction to *Hearts on Fire* I suggested you can use the pain to "break on through to the other side." There is no way out of the pain

---

[3] When I say "I" or "you" it means the nervous system, because the "I" or "you" is a by-product of the Nervous System.

other than through it. Hopefully, if you are willing to go with it with awareness, using the tools to navigate through the pain without the intention of getting rid of it, you might go Beyond.

QUANTUM PSYCHOLOGY PRINCIPLE:
Taking the False Core apart without the intention of getting rid of it, healing it, transforming it, etc., is like taking the label off and having it as energy without the intention of getting rid of it.

Focus on the False Core. Look at it without trying to get rid of it. *Do it to do it.* Take off the label and see what happens. There are no guarantees or promises. "I" don't know what's going to happen.

There was a lady from Florida who had like 25 years of therapy. Her False Core was "I don't exist." Now, given that she has had so much therapy, I am not going to go over her mother and father stuff. But the bottom line was, I knew she was going to *hold on* for dear life to the images of mom and dad and what "he said/she said." She did this as a way to avoid feeling her False Core since it was better to "hold on" to this childhood story than to deal with "I don't exist." In fact, the only way she existed was to hold on to her images. I did not focus on anything except, "You will do anything to resist your False Core." At this point, she began to "own her False Core."

The general aim is to return (discard) the False Core-False Self to whoever, oftentimes mom (lineage) or dad (lineage) modeled the False Core-False Self. Usually at this point clients can see how they absorbed it from them. Basically you are giving back (discarding) what you took on (fused with) and "getting it is not you." The lineage is deep and within the body. It should be noted there is no blame or shame in this process at all. It is more like taking off an old shirt you thought was you. Oftentimes people think and feel mom and dad live in their body or they can feel it is in or under their skin.

Or as Cole Porter wrote and Sinatra crooned:

I've got you under my skin. [So take off your skin.]
I've got you deep in the heart of me.

So deep in my heart that you're really a part of me,
I've got you under my skin.
I've tried so—not to give in. . . .

QUANTUM PSYCHOLOGY PRINCIPLE:

The problem is you (nervous system) are always re-creating your False Core and, at the same time, trying to get rid of it or compensate for it through the False Self.

The nervous system labels experiences and also decides what these experiences mean. To appreciate this fully, let's take the feeling of anger and imagine if we could slow down the process for a minute. It might look like this: First, you feel bodily sensations. Second, the nervous system labels these sensations as angry. Third, the nervous system (brain) wants to get rid of the anger. Why? Because children are taught by their parents that anger is bad. Fourth, the nervous system's survival mechanism kicks in and says, "I must get rid of the anger in order to survive with my parents."

As you can see, the False Core-False Self is like a live entity with its own life. It has a structure, purpose, and force of its own, its purpose being survival. It will continually reinforce itself and not allow anything which threatens its existence. If something completely opposite is let in, from the point of view of the False Core, it is mistakenly seen like offering kryptonite to Superman and, hence, resisted.

FUSIONS OF THE FALSE CORE

The False Core can be fused in several ways. It can be fused and triggered by thoughts, feelings, biological functions, **ESSENCE**, **I AM**, etc. For example, a woman I worked with at a workshop had a feeling which triggered the False Core of "worthless." Whenever *any* feeling came up, she felt *worthless*. The feeling was automatic (see demonstration, Chapter VII).

Or if you have a False Core of "I don't exist" (to be discussed later), the False Core is trying to prove *existence* and *nonexistence* simultaneously. Because they believe "I'm nothing" and "I don't exist," they try to accumulate thoughts and, in doing so, become *over* thinkers. But you will see them frequently in meditation groups where they are trying to *get rid* of thoughts. So they have fused together "thoughts equals existence," but thoughts *and* existence are the opposite of non-existence. Therefore, there is always a push-pull, a dissonance, an ambulance which operates outside of their awareness. They accumulate thoughts and simultaneously, they want to be thoughtless.

You can fuse together that thoughts equal the False Core or feelings equal the False Core and their interruption or conclusion about the external world triggers the False Core. I have had several so-called spiritual people say "I don't want to have anything to do with the external world." "I don't like the external world." "I wish the external world would go away." I asked them, "What does the external world mean to you?" And they said, "The external world means, *I have to do something*." Okay, if you fuse together that the external world equals doing, what gets created, and what does not get created, etc.

EXERCISE
FUSION OF THE FALSE CORE (PAIR UP)

Summary of Six Step Defusion Process:[4]

The False Core "should be un-fused from all dimensions. In this process run it with 1) external, 2) thoughts, 3) emotions, 4) biology, 5) **ESSENCE** and 6) **I AM**.

Question I: If you fuse together (one of the dimensions) and the False Core, what do you create? Or what gets created?

Question II: If you fuse together (one of the dimensions) and the False Core, what do you not create? Or what does not get created?

---

[4]The application of this process can be seen in *The Tao of Chaos*.

Question III: If you fuse together (one of the dimensions) and the False Core, what do you resist? Or what gets resisted?

Question IV: If you separate (one of the dimensions) and the False Core, what occurs?

Question V: If you separate (one of the dimensions) and the False Core, what does not occur?

Question VI: If you separate (one of the dimensions) and the False Core, what, *if anything*, gets resisted?

THE BIOLOGICAL DIMENSION

You can also fuse together the False Core with biology, placing it on a biological function so that it continues to be on automatic. You can fuse the False Core to your eye blink, your heart beat, your muscular contraction or your breathing. For example, if somebody threatens your False Core, you might feel your biology collapse, contract and tighten to protect your False Core. My mother used to say, "From the moment I open my eyes to the moment I close them, I get no peace." She had the belief structure called "I get no peace" attached to her eye blink which fused the biological function of eye blink and False Core. The False Core can be fused with breathing. That's why, if you go to a breath therapist and they have you breathe long enough, memories and emotions come up.

What's interesting is that in many meditation practices, you shut down the biological aspect of the False Core. Meditation usually requires that you close your eyes (no blink) and watch your breath or heart beat. Naturally, when you *focus* like this, the heart and breathing pattern slow down yielding relaxation. Then, of course, you open your eyes and breathe normally and everything comes back again. Your mind comes back because the False Core-False Self is attached to a biological function and is triggered by it. Let's say a little girl was sexually abused and her perpetrator says, "You're bad, you should be ashamed of yourself for seducing me." She then fuses sexual feelings with the perpetrator even though he leaves her after the act. So now she has sexual sensations or body feelings fused with the False Core, which is fused with sex. Years later when sexuality comes up, so does the False Core.

THE ASSOCIATIONAL TRANCE REVISITED

As mentioned one of the deepest tendencies of the mind is the *Associational Trance* which creates associations (i.e., past equals present, past equals future). In other words, the present situation is the *same* as the past. The purpose of the Associational Trance is to organize and generalize events into earlier similar events in order to *avoid* the perceived chaos of *new* and *different* experiences. It assumes, through the creation of associational patterns, that events are not new but associated and related to one another, in the form of a pattern. The *breaking of the associational trance leads to the realization of the saying, "You never put your foot in the same river twice."*

Quantum Psychology wants to *cut* this associational trance, network or chain. If you see the False Core for what it is, it's very easy to deal with. But when your False Core is wound up with your biology, your thoughts and all of the traumas in the Associational Trance, everything is collapsed and fused together. When this is the case, you cannot experience the False Core as the False Core. For example, if you are in an association of the False Core, "I am "inadequate," and it gets associated with"I'm stupid," "I don't understand," "My mother never loved me," "My second grade teacher yelled at me," "My boyfriend told me I'm stupid"—these are associations. By cutting this associational trance, you can experience the False Core as the False Core which is just a lens or point of view.

When the narcissistic wound happens, the False Core is formed and ultimately develops a life of its own, it blames the mislabeled spaciousness of **ESSENCE** for the separation and mislabels the spaciousness of **ESSENCE** as emptiness (as in a lack) and not-me. It then attributes qualities which it perceives as bad to the mislabeled spaciousness. The False Core then imagines the emptiness will re-enact the shock of the Realization of Separation. For example, if an infant is angry and mom separates further, he or she attributes the anger to the mis-labeled emptiness which was **ESSENCE** (with a *lack* of label placed on it), and then decides that this (anger) is "not me" and projects it onto others. This projective identification reinforces the disowning of the attribute (in this case, anger), thus reinforcing and re-enactment of the narcissistic wound. *All of the unwanted or unaccepted attributes which mom does not want get placed on the mis-labeled spaciousness of ESSENCE, now emptiness (as in a lack) and become the shadow or unconscious mind* which can get fused with **ESSENCE**.

This tendency does not stop with painful emotions; even pleasant feelings such as love and "intimacy" can trigger the shock of the Realization of Separation. Why? Because since **ESSENCE-I AM**, with its essential qualities like love and unity, etc., is also fused with the Realization of Separation, they (the essential qualities), when they occur within a relationship, unfortunately bring up the associated pain which is fused with the Realization of Separation and the wound. For example, in a relationship when love and vulnerability come up they can get associated and fused with the shock of the Realization of Separation. When this occurs you become certain you will be abandoned just like mom abandoned you.

All the accepted attributes continue to be manifested by the conscious mind (False Self). The repressed and denied attributes have a force and a life of their own, they continue to be "acted out" unknowingly (False Core). And so the shock of the Realization of Separation and the accompanying narcissistic wound continue to be re-enacted again and again.

ASK YOURSELF THESE QUESTIONS:

Although this is premature, as the False Cores themselves have not been officially named, this exercise can lead us to what will come; i.e., the False Core-False Self and its relationship to the shock of the Realization of Separation and the narcissistic wound.

Exercise 1   In order to not feel X (the False Core), I won't (*fill in the blank*), for example, have certain thoughts or feelings. Repeat until nothing more emerges.

Exercise 2 In order to not feel the False Core, I will (*fill in the blank*).). Repeat until nothing more emerges.

Exercise 3 When I feel the False Core, I will (*fill in the blank*), for example, make up a story to justify its importance, etc.). Repeat until nothing more emerges.

BLAMING ESSENCE-I AM

ESSENCE-I AM is blamed because **ESSENCE** is mislabeled as empty (as in a lack) and blamed for the shock of the Realization of Separation. In other words, the reason that I'm separate is because **ESSENCE-I AM** is so (*fill in the False Core*). I blame **ESSENCE-I AM** and this mis-perceived emptiness or lack for the separation. That is why emptiness is *so* resisted at a psychological level because emptiness is not emptiness itself or spaciousness—but rather it is associated and fused with a lack; like I am, *less than* etc. In this way, since the only reference point prior to the separation is **ESSENCE-I AM**, therefore, it has got to be the fault of **ESSENCE-I AM**. When this occurs, you start splitting off from **ESSENCE-I AM**. All "bad" attributes are ascribed to **ESSENCE-I AM** and it becomes the disowned shadow. The mis-perceived emptiness (lack) of **ESSENCE-I AM** once labeled is then compensated for.

ESSENCE-I AM is mistaken for a lack, and blamed and labeled. The False Core becomes the concluded root cause or the reason I'm separate from my mother (i.e., because this (mis)perceived empty is a lack and must mean (*fill in the blank*), i.e., "There's something wrong with me," etc.). And for that reason, I must overcome and compensate for the label (False Core) I placed on the (mis)perceived emptiness as a lack which I do not know or understand is **ESSENCE-I AM** (see Volume III).

QUANTUM PSYCHOLOGY PRINCIPLE:
The label placed on the (mis)labeled spaciousness of **ESSENCE** now becomes emptiness (as in a lack) covers **ESSENCE-I AM.**

QUANTUM PSYCHOLOGY PRINCIPLE:
It should be noted that it is not **EMPTINESS** that is resisted—rather the (mis)perceived emptiness or lack label which is placed on the spaciousness of **ESSENCE**, once delabeled the emptiness (as in a lack) is realized as an Essential Quality of Spaciousness.

ESSENCE'S LABEL
The label placed on **ESSENCE** *becomes the shadow*. The shadow, in turn, becomes an archetypical explanation as it becomes more disowned. When I disown and blame **ESSENCE**, this causes me to look for myself (**ESSENCE**) at the same time that I disown myself (**ESSENCE**). The problem is, I can't find myself because I've labeled **ESSENCE** (myself), blamed **ESSENCE** (myself) disowned **ESSENCE**, (myself) and split off from **ESSENCE** (myself) which I now imagine is a shadow side of me. **ESSENCE** is disowned and blamed for all these bad qualities, and for my imagined fall from union with mom which is later spiritualized as the fall from grace or God

The False Core and its attributes are fused with the emptiness (bad label) and denied as Not-me. It is projected onto others, and made into the unconscious mind. This is done unknowingly because the pereceived emptiness or lack is Not-Me and is blamed for the narcissistic wound.

ARCHETYPES
The shadow as an archetype are denied attributes which are placed or trance-ferred on **ESSENCE, I AM, NOT-I-I,** etc., and

Chapter II

which have been deemed bad by the external object (mom). Thus, the implicit/explicit message from mom/dad/society, etc., is seen through the eyes of the False Core as, the way to merge is to disown or project these attributes on others. This creates the shadow. So in order to survive, you must unconsciously project unwanted aspects onto another. Archetypes (see Volume III) are often spiritualizations and mystifications assigned to resist the Realization of the Separation between mom and infant and to create a story about "how to merge." In other words, the quest for the Holy Grail, is a search for the secret of how to heal or merge after the Realization of Separation. This applies to the search for the realization of unity or enlightenment.

The False Core and all its attributes are fused with the now mis-perceived emptiness and lack and, hence, denied, projected onto others, and made into the unconscious mind. This is done because the emptiness now called the not-me is blamed for the narcissistic wound.

The "fall of man," "original sin" and "redemption" are metaphors for the Realization of Separation and its healing. Since parents are trance-ferred onto God, these are displaced, spiritualized concepts caused by the shock of the Realization of Separation.

I must hide my defect, the reason (False Core) that I am separate. I must (False Self) repent and overcome it or heal it so that I can go back and merge with mom/God. This search has now become a psycho-spiritual path to unity, trance-ferred and spiritualized from the shock of the Realization of Separation. Parental rules are trance-ferred onto God and become spiritualized, thus becoming the **WAY** or means to unity (merger with Mom).

THE MOVEMENTS OF THE FALSE SELF

The mind and its associations, ruled and masked by the False Self, move to avoid the False Core. Quantum Psychology calls these movements the five R's: 1) To *resist* the False Core by trying to overcome it. 2) To *reinforce* the False Core. 3) To *reenact* the False Core. 4) To *recreate* the False Core. 5) To try to *resolve* the False Core through healing or transforming it by *over-compensating* with (*fill in the blank*).

If my False Core is "I am inadequate," the False Self can try to *resist* by trying to prove adequacy in some way or avoiding inadequacy. The False Self can *reinforce* it when I try to overcome it. The False Self can *reenact* it by setting myself up for situations where I feel inadequate. The False Self can try to *recreate* it. You (nervous system) will always recreate your False Core in every situation as a repetition compulsion (to get it right) or, most insidious of all the False Selfs, you could try to *resolve* the False Core by trying to heal or transform the False Core by being overly adequate (more healthy) which only re-enforces the holographic nature of the False Core-False Self relationship further.

Remember if you think the False Core is you and you get into a "healing process," or become a healer in some form, since the False Core and the healer (False Self) are part of the same structure, this False Core must always be maintained in order for the healer (False Self) to do his job (to be discussed later). The movements of the Enneagram are the movements that the mind (False Self) uses to avoid the False Core. They mirror the movements of the mind. There can be numerous ways the False Self preserves, the manifestation of the False Core. In the above example, you always prove the (False Core) *I am inadequate*. The rest—what happened or did not happen, etc.—just becomes a made-up story to justify the outcome, i.e., being left with your False Core.

DOES THE FALSE CORE COME BACK AFTER IT IS "WORKED-OUT?"

Under stress the False Core will come back, but over time you will realize it right away and immediately discard it rather than acting it out. A student once asked Nisargadetta Maharaj, "Does anything ever come up for you?" "Every so often something comes up," he said, "but I *immediately* realize it's not me and discard it." So, it's important to know about the False Self and be able to *trace* every thought, emotion, and fantasy, etc. back to the False Core since they all originate from, and are feeble attempts to resist, the False Core.

Then go into the non-verbal **I AM** prior to the False Core and see the False Core and its associations as not you (see Volume III, Chapter, **I AM**). I was once scheduled to give a workshop in Nevada and I had been driving on the road for hours. I was tired and my body was all stressed out. At the motel, I got into the shower and suddenly my False Core arose. I looked at it and it disappeared. Again it arose and again I noticed it and it disappeared. This happened three times.

And so, the more the False Core loses its impact on you, the more your psychology will no longer impact you. *From the outside, other people might see you behaving in a certain way and interpret this as you being in your False Core. But internally and subjectively, you will remain unaffected. There is an old saying in Raja Yoga— "when a pickpocket sees a saint, they only see their pockets." In this way, if you are in your False Core-False Self, you only see your own and other people's False Core-False Self.*

BEGINNING THE FALSE CORE DECONSTRUCTION PROCESS

Step 1 - Acknowledge both sides, i.e., The False Core-False Self.
Step 2 - Own it.
Step 3 - Un-own it, Get it is not you.
Step 4 - Dismantle.

CHAPTER III
THE FALSE SELF

The idea of a **False Self** identity is not new. I first came across the term thirty years ago in the work of D.W. Winnecott in a beginning class on Developmental Psychology. Developmental psychology views the **False Self** as a buffer against possible harm to what is called the "real self."

Before we continue to discuss what the real self is, let us look at what Quantun Psychology suggests it is not. According to Heinz Kohut, the father of Self Psychology, the real self is a combination of genetics and biology coupled with positive internalizations of the parents. In other words, in self or ego psychology, the real self is a combination of your inherited genetics and the positive fused identities, or positive parts of yourself taken on from others. Simply put, Kohut is saying that the self is the body or genetic somatic self and the positive parts of mom and/or dad and significant others that you took on and which you now call yourself.

WHAT MAKES QUANTUM PSYCHOLOGY DIFFERENT FROM EGO PSYCHOLOGY OR OBJECT RELATIONS?

In ego psychology and object relations, the self or the ego is what remains after the natural drive impulses such as sex, food, etc., are regulated by outside people like mom and dad. This outside regulation later becomes an internalized mom and dad in what has traditionally been called the superego. The impulses of the body are called the id, and the ego is that which comes out of this inherent conflict between internal drives and impulses and the outside regulators. As mentioned earlier, according to Kohut, the real self is a combination of genetics plus the positive internalization of the parents.

Psychoanalytic theory presupposes developmental stages which, when they are not completed, cause a diminishing in the development of the natural real self. Most schools of psychotherapy from Freud to ego psychology, from Jungian therapy to object relations, attempt to repair and "put in" or "correct" what happened in your chlidhood. From transactional analysis to Ericksonian work, attempts have been made to heal, re-frame, re-associate, re-decide and re-organize in present time, what happened in your childhood.

This presupposition of repairing and "righting" the "wrong" and then being left with a healthy self bears a "scary" resemblance to the origin of repetition compulsion; going back in time (in therapy) and doing, correcting, reorganizing, etc. In short, righting the wrong. *This is where Quantum Psychology and most forms of psychology differ.*

QUANTUM PSYCHOLOGY VS. EGO PSYCHOLOGY

THE REAL SELF

In Quantum Psychology, the real self is *not* the somatic self plus positive internalization. Your deepest "individual" real self, Quantum Psychology calls **ESSENCE** (Essential Core) or the **I AM**. It is who you were **PRIOR TO** you taking-on any "positive" or "negative" internalizations from your parents. **ESSENCE** or **I AM** can be realized by discovering who you were *prior* to you taking-on any programming or internalization. The **I AM** even precedes the essential qualities of **ESSENCE** and it can be said that the **I AM** is the no-state state, a quality-less state *prior* to essential qualities.

In other words, the *I AM is the essence of ESSENCE* (see Volume III). To experience the **I AM**, let your eyes close and without using your thoughts, memory, emotions, associations or perceptions, notice **WHO YOU ARE**. In a word that no-state state *prior* to thoughts, feelings, emotions, associations or perceptions, is who you were before you took on any programming. This experience is impossible to fully appreciate through thought or understanding. In Quantum Psychology training programs, however, **I AM** is very available and integratable.

Quantum Psychology, because it follows Nisargadatta Maharaj, stresses that this "you" *prior* to your programming is the pointer and direction to the discovery of **WHO YOU ARE**. Unfortunately for us all, the identities of external others, coming from the external dimension and later internalized, inhibit the outward motion of energy, thus driving the energy upward into thoughts and trances. Therefore, to reach and stabilize in the Awareness of **I AM,** dismantling the False Core which hides the **I AM** and **ESSENCE** is suggested. *The False Self, which Jung called the persona hides, tries to heal, or overcompensate for the False Core, also must be dismantled.*

What you are left with then is a real somatic body with feelings, thoughts, etc., an animal nature, **ESSENCE** with essential qualities and the **I AM** without the associational trance of the imagined "past" projected onto the present and imagined "future

DISMANTLING VS. REPAIR

Since the False Self, or *persona*, which hides the False Core is not you but was created or taken on by you to survive, it is important for three reasons to not repair the ego, but to dismantle it. First, it is easy to fall into the False Self's trap of trying to resolve, heal or transform the False Core. This places you on the endless False Self→False Core→False Self→False Core→False Self→False Core merry-go-round. Second, the False Core is a concept—it is not you, why try to repair what is not you, and third, you cannot heal a False concept.

What you are left with is neither inner voices nor positive/negative internalization. You are left with just yourself. The non-verb **I AM**.

The Way of the Human • The False Core and the False Self

QUANTUM PSYCHOLOGY PRINCIPLE:
Any concept that was preceded by a False concept must also be False.

In this way, the False Core concept (conclusion) precedes and begets the False Self. This means that the False Self's attempt to heal or transform the False Core must ultimately be fruitless. Why? Because the *False Self's conclusion of healing or transforming the False Core is based on, rooted in, and driven by the false premise of the False Core.*

THE FALSE SELF
The False Self is a compensation to defend against the False Core. It acts as a buffer and is the way we present ourselves to the world, i.e., our socially acceptable masks. In this way, we hope that our needs will be met. The False Self is created when a child is hungry and screams for mommy. Mommy lets the child know the socially acceptable ways to get fed. The child then adapts his or her behavior and creates a socially acceptable identity, persona, or mask of "appropriate" and "acceptable" ways to get their needs met. Compulsive, unsatifiable wants are characterized by, "It's never enough." This happens when our ummet biological needs are substituted for socially acceptable, "psychological" wants.

This False Self is an adaptive identity which becomes automatic, a False Self identity we operate out of in order to survive. Unfortunately, we forget the identity is fake, a made-up creation. Years later we wonder why we feel alienated and misunderstood, not realizing it is the result of being stuck in the False Core-False Self complex.

Take, for example, a sexual relationship where we can't say what we need because it might be considered crude or uncool. Instead, we pretend, we create a fantasy and hide our true somatic feelings in the vain hope that we'll get what we want. To paraphrase Marlon Brando:

Chapter III

You're having dinner at someone's apartment, it's the second or third date, and sitting on the table is a hippopotamus which says, "I wonder if we're going to have sex tonight?" And both of you are pretending the hippopotamus is not there.

This is the nature of the False Self—to buffer, pretend and defend the impulses, feelings and drives of the somatic self and to hide the False Core. It forces us to split off from our somatic and animal self creating an image and hiding our true feelings.

Recall a time when you felt angry or upset and put up a false image to hide it from the world. Or a time when you felt sexually turned on and pretended you weren't by creating an air of indifference. This is the False Self in action.

We deny our anger and our sexual feelings in favor of a False Self (image) which we fall in love with and try to get others to do the same. But this is ultimately unsatisfying because *we KNOW it is FALSE*, and so any "love" we receive by or through it can never "be taken in" because it was received under *FALSE* pretenses. In this way, the False Self not only defends against the False Core but it denies our bodily sensations and animal nature.

Understand that, the False Core-False Self is not you. This is why "acting out" of an identity has to leave you feeling alienated, lonely, misunderstood, and separate. The only way to really feel unity with another is **ESSENCE** to **ESSENCE** or **I AM** to **I AM**. In order to do this, we must first discover and dismantle our false selves (compensating Identities) and our False Core.

IDENTITIES UNDER STRESS

Identities are oftentimes taken on under high stress. Take the example of a child being abused by Dad. Under this kind of stress, the child will take-on Dad's identity. In this way, the child will develop two identities, which act as lenses by which he views and experiences the world. In this instance, we could call abusive Dad, Identity One, and the victimized child, Identity Two.

The idea that identities come in pairs of opposites or polarities is not new. For hundreds of years, the *Yoga Psychology of Patangali* known *as Patangali's Yoga Sutras*, has spent much time just on this. Identities or states come in oppositions and are generally labeled as "bad" or "good." For the most part, the discovery of *WHO YOU ARE* is really the discovery of *WHO YOU ARE NOT*. All of us take on, fuse with, and become part of significant others in our life. "I" have found over the years that many identities that are "taken on" come in patterns and can be seen in particular combinations and in the following ways:

FUSED IDENTITY COMBINATIONS

I.   I-dentity #1 Mom   I-dentity #2 Mom
In this combination, the child takes on two (sides) identities of mom. For example, a child may fuse with identity #1, depressed mom, and then fuse with the second identity #2, supermom.

II.   I-dentity #1 Dad #1   I-dentity #2 Dad
The child takes on two sides (identities) of dad. Identity #1 could be alcoholic dad which the child fuses with; and Identity #2, the over-achieving dad.

III.   I-dentity #1 a Mom   I-dentity #2 a Dad
Here the child takes on two identities, one of mom and one of dad. One identity might be mom telling dad, "Don't drink." The second identity could be dad saying, "Drink."

IV.   I-dentity #1 Mom and I-dentity #2 a created response to Mom identity.
In this example, the child might take on and fuse with a strict mom identity (#1). Then, to resist mom, they might create a response to the mom identity (#2) called, "I'm going to do what I want."

Chapter III

V. I-dentity #1 Dad   I-dentity #2 a created response to Dad identity

Here, the child takes on one side of dad and then creates a response to dad. Let's say dad is a fascist. In response to this, the child fuses with fascist dad and also creates a communist identity. Therefore, the child has two identities—fascist dad and communist son.

VI. I-dentity #1 a bad identity   I-dentity #2 a good-overcompensating identity

In this example the child is given an identity called bad or stupid, and then creates an overcompensating identity of trying to prove their goodness or intelligence. Another example might be an unworthy identity. To overcompensate, a person will create an identity which is always trying to prove its worth.

THE FALSE CORE PROCESSES

POINTS TO REMEMBER

1. The first identity is the strongest. In other words, if we use the example above, the first taken-on identity called "I'm worthless," is the strongest. So the continual attempt to prove one's worth is an attempt to overcome worthlessness. But nothing can be overcome because all attempts to "prove worth" are *driven* by worthlessness. For this reason, no matter how many times the person succeeds in life—deep down, they feel bad about themselves.
2. The "You" (**I AM**) was there and is *prior* to any identities. And while the identity called worthless (used in the above example) is there and after it leaves, the same **I AM** is there. *You cannot be and are not your identities*. Often in therapy, to get a client to appreciate that they were there *prior to* their identity, I say to them, "Tell me the difference between you and this image, feeling or thought called (*fill in the blank*)."
3. The "I" you call "you" is part of the False Core-False Self complex. When it disappears there is No-I.

Nisargadetta Maharaj used to say, "Who came first, you or this 'I' (*I*-dentity)?" Or find out who (what) "I" wants to know. Stay *prior to* your last thought.

Unlike other psychological schools or "spiritual" systems, Quantum Psychology does not suggest that some Identities are good, healthy or more virtuous while others are unhealthy or bad. Quantum Psychology's goal is for people to find out who they were and are *prior* to their False Core-False Self. This is done by dismantling the False Core-False Self and other identities so that we can become aware of **ESSENCE**, the **I AM** and beyond.

THE WAY OF THE HUMAN

Reviewing our Steps: The Tao of Chaos Revisited:

It is important at this juncture to trace the steps from **ESSENCE** to psychosomatic disorder. Let us begin, (see *The Tao of Chaos*) with **ESSENCE** which has neither observer nor observed. When the body organizes around the spaciousness of **I AM** or **ESSENCE**, the body continues to have its biological needs. After the shock of Realization of Separation some needs are met while others remain frustrated and ignored. To combat this contradiction between **I AM** or **ESSENCE** and the frustration and pain of unmet needs, comes a major dissociation: the emergence or creation by the nervous system of *the observer which is an identity*.[1] The observer is part of the Identity which begins by seeing things, namely, the **I AM** and **ESSENCE** as "not me." Hence, the observer develops self-consciousness, and dualistic perception (i.e., becomes aware of itself). For example, "You are there," "I am here, or "I can observe my thoughts." In this way, not only are others seen as "not me" but **I**

---

[1] Neurologically speaking, the observer comes after the "shock" on which it places the False Core label. Thus, the observer is a by-product of the newer "function" of the brain. Because the newer brain (cerebral cortex) came later in evolution, it is physically higher than lower brain function. It seems feasible as a theory but not a proven fact that the "higher" parts of the brain label and judge the lower parts as bad or less than itself. If this is true, then the brain itself would label the body, emotions and sensations, coming as they do from these "lower" parts as less than and less "spiritual."

Chapter III

AM and **ESSENCE** might also be seen in this way and, thus, as something to be a-voided.[2]

Once the observer-identity is formed, it (mis)labels the Essential quality of the spaciousness of the **I AM** or **ESSENCE**, around which the body has organized, as emptiness, hence, it is viewed in some derogatory, less than, or as some kind of a lack. For example, "There's something wrong with me," "I'm worthless," etc. For this reason, **I AM** or **ESSENCE** begins to be seen as a lack of something and the label affixed to it becomes the False Core. Quantum Psychology has found that the False Core is the most basic label placed on the spaciousness of **I AM** and **ESSENCE** after the shock of the Realization of Separation, which we then resist because **I AM** or **ESSENCE** is also fused with the shock.

The False Core (Driver) is the derogatory label which gets placed on **I AM** or **ESSENCE** which is then resisted. At the same time, this False Core becomes the lens through which "you" view the world, "yourself" and how you imagine the world views "you."

When the observer labels **I AM** or **ESSENCE** as unworthy, for example, all experiences are funneled through this lens of "unworthiness." Events are taken in, experienced and digested so as to reinforce *it* (the False Core concept). Thus the spaciousness of **I AM** or **ESSENCE** has been (mis)labeled as unworthy, In this way, all experiences reinforce this observer-identity of "unworthy," with not too much regard to the external world. In short, the observer-identity of unworthy becomes the filter of most experiences of the personality. In this way, the observer/identity combination becomes your False Core and the way you defend against it (i.e., your mask, whether it's psychologized or spiritualized) becomes the False Self.

To clarify, the observer is part of the False Core-False Self complex. **I AM** and **ESSENCE** are beyond the observer. The observer's job is to observe and perceive through the lens of the False Core and False Self. In this way, if you are in observation (observer mode), the False Core acts as a lens through which the observing of the observer takes place. To stay stuck in observer mode means your awareness can never stabilize in **I AM**, **ESSENCE**, etc. Therefore,

---

[2] In **I AM** or **ESSENCE**, observation means observation with no object. Gurdjieff called this Objective Consciousness.

ultimately, to stabilize in the next dimension of your human nature—**I AM** or **ESSENCE**—you must go beyond the observer, the False Core Driver and the False Self Compensator. Since they are one unit, as we go on we will call them the Observer-False Core-False Self Complex.

This "not me," along with the (mis)label affixed to **ESSENCE** or **I AM** and the "you" you think you are, become the False Core and False Self. This tendency forces the observer to unknowingly fixate attention on the way **ESSENCE** or **I AM** is (mis)labeled. In this way, you wind up thinking you are your False Core and the compensating Identities called the False Self. When these are dismantled, there is no you.[3]

To combat this False Core, which the observer cannot help but look through, the observer develops an overcompensating identity—a False Self. For example, if **ESSENCE** or **I AM** is (mis)labeled bad in some way by the observer, then you cannot help but fixate unknowingly on this label, thus creating a False Self identity of proving you are not bad but are good to over-compensate for it. The False Core is placed on **ESSENCE** or **I AM** which justifies the imagined reason for the Realization of Separation (see next Chapter). When the observer fixates on it unknowingly, it is unable to escape except through over-compensation.

According to Quantum Psychology, these labels placed on the "personal" spaciousness of **ESSENCE** or **I AM** create (mis)perception and (mis)information and, hence, the reasons and justifications for the False Core-False Self. The False Core-False Self grabs attention and forces attention to remain fixated, thus creating an entire internal psychological world of Maya or illusion. It must be kept in mind that the False Core label was designated by the observer, and that the observer at this level is a by-product of, or was created by, the nervous system through biological deprivation and shock and which, in turn, created a dissociation or a splitting-off from the animal dimension.

To illustrate this splitting off, let us take the Evangelist Jimmy Swaggart who *seems* to be using two of Oscar Ichazo's identities, the

---

[3]To be discussed in Volume III. For now, everything you think or imagine yourself to be originates from the False Core or False Self. Which is not you. Therefore, ultimately, there is no you.

hedonist and the puritan. Swaggart will *split off* from being a hedonist and preach (puritan identity), thus projecting hedonism onto his flock. Later he will *split off* from his puritan identity and again become the hedonist. When he is being the puritan identity, the hedonist is seen as "not me"; then the hedonist is spiritualized, "The devil made me do it" (have sex). When he is being the hedonist identity, he openly pursues prostitutes, and the puritan is seen as "not me." The extremes of his behavior demonstrate the biological repression of his animal nature (sexuality)[4] which forces him to *split-off* and form the puritan.

What will be discussed in greater detail later but bears mentioning now is that Quantum Psychology does not separate body from mind. Rather, Quantum Psychology says the body is the mind and the mind is the body.

BETRAYAL

One of the major causes for the creation of False Self identities is betrayal. It is a form of chaos which is caused when someone we trust is unfaithful, disloyal, or deceives us in some way. Identities are formed in an attempt to handle or overcome this crisis.

The shock of the Realization can be (mis)perceived and (mis)labeled as a betrayal. This (mis)perception leads to other (mis)perceptions of betrayal where there might not be any and, thus, acts as a defense against this "imagined betrayal" simultaneously. Please note, the *Realization of Separation is not a betrayal, it is a natural process*, what it was "decided" it means, defends against the shock of the Realization of Separation, and leads to misperceiving external reality years later.

The common definitions of betrayal, although easy to understand, left me without appreciating its origins, its nature, its motivation, and most important, its impact on the individual.

Let us first look at a new definition of betrayal in the light of the developing individual.

---

[4]Sexuality is seen as biological and part of survival, i.e., *the survival of the species*.

To Betray Another:
1. To betray: to ask someone to be something other than what she/he actually is.
2. To betray: to present yourself to another other than the way you are.
3. To betray: to mislead by adding, deleting, withholding or changing information in your presentation to another, specifically in an attempt to control them and control their reactions, or get from them what you want.

To Betray Yourself:
1. To betray yourself: to tell lies or stories to yourself about the way you are.
2. To betray yourself: to tell lies or stories to yourself about the way someone else is.
3. To betray yourself: to tell lies or stories to yourself to justify the way you treat another.
4. To betray yourself: to tell lies or stories to yourself to justify your position or viewpoint.

THE ORIGIN OF BETRAYAL

The origin of betrayal lies deep within our personal history. Each of us has experienced it. The experience is rooted in survival because during the Realization of Separation we imagine and (mis)perceive it as betrayal and so feel our survival is in jeopardy. It arises because there is an implicit belief of on-going merger with Mom. Since this cannot and does not occur, due to the shock of the Relaization of Separation, we feel betrayed. In this way, the label of betrayed is fused with the shock of the Realization of Separation and its outcome the narcissistic wound. In this way, we attempt to avoid betrayal by betraying oneself and others.

Because of this the nervous system *scans* the environment to read for possible betrayal, or avoids and defends against possible betrayal as part of the survival mechanism. The problem is that the scanning-searching device is not in present time—but remains in the past time shock and defense against the shock. In this way the perception is inaccurate since there was no betrayal during the shock of the Realization of Separation; it was a natural process. The betrayal is a (mis)perceived betrayal. In this way the scanning-searching overgeneralizing mechanisms of the nervous system which provide for "our" perception betrays our experience of present-time reality. In this way, *we can only betray ourselves and others.*

For example, a child is asked to betray his true feelings, wants, desires, or thoughts to get what he needs from his parents. Parents betray themselves by denying or not expressing their feelings toward their children, pretending that they care, or that they are listening when they are not. *The very act of pretending and faking feelings is betrayal.* And in this society, pretending is the rule not the exception. Recall the old expression, "Fake it until we make it." The unfortunate part is that in our family system and culture, betrayal is socially acceptable. The feeling of being betrayed is so accepted that it has become part of our psychology and, more importantly, of our society.

We have accepted betrayal in its myriad forms as the way life is. We expect and accept betrayal in our business dealings. We expect and accept betrayal in our relationships. We expect and accept betrayal from our politicians, religious leaders, teachers, etc.

We betray others and ourselves when we ask them to be other than the way they are. This helps to create and re-enforce the narcissistic wound or injury. We objectify others and lose the fact that they are human, by not seeing them as they are in present time. Instead, we see our intrapsychic picture of them. We then try to manipulate them so that they conform to the picture in our mind. This is especially true in our close relationships when we ask a person to change rather than allowing them to be who they are. For example, I recently saw a woman for therapy who was always trying to get her husband to change. I said to her, "Obviously, you do not like your husband." She said, "What do you mean?" I said, "If you liked him the way he is, you wouldn't try to get him to change. You don't like him the way he is, you like him the way you want him to be." And we betray ourselves by trying to be something other than who we are.

THE IMPACT OF BETRAYAL

What is the impact of betrayal?
1. The formation of a False Self;
2. the objectification of another;
3. the loss of yourself and your present-time body experience;
4. a deep unacknowledged and unaccepted grief that underlies and pervades the basic fabric of our bio-psychology;[5]
5. the creation of a psychological philosophy in our society which accepts and supports betrayal, e.g., "business is business";
6. the participation in a world of objectification of others and ultimately of ourselves as living, breathing human beings.

In the end we wind up betraying our bio-psychology. Once we betray and objectify ourselves and others in order to get immediate gratification, we relate out of an identity—not out of **ESSENCE** or **I AM.** In this way, we lose our humanity, and hence, our connection to the underlying unity. Betrayal is a major ingredient in our creation of identities and our philosophical rationalization makes it palatable. But betrayal is not *The Way of the Human*, it is the way of the automaton.

CONCLUSION

Quantum Psychology views **I AM** or **ESSENCE** as the spaciousness that the body organizes around. It is (mis)perceived, blamed and (mis)labeled during the shock of the Realization of Separation as well as the pain of unmet biological needs. This label becomes the reason and is the False Core. The observer then appears with identities to help it in its task of enabling the physical body to survive.

---

[5]Dr. Ernest Rossi needs to be acknowledged for his use of the word "psycho-biology." I have reversed the order to reinforce that the biology came first and that psychology is a by-product of biology.

Chapter III

Compensating Identities are the False Self. The False Core is the driver of the False Self. In other words, the False Self was created to handle the False Core. No False Core, no False Self. **ESSENCE** and **I AM** are the deepest sense of a being. To quote Gurdjieff, "**ESSENCE** is what you were born with; personality [identities] is what you acquire."

Quantum Psychology sees that 1) the observer is part of the personality; 2) there is a different observer for each experience; and 3) to dismantle identities and each corresponding observer is important because it shrinks awareness. In this way **ESSENCE** and **I AM** can "move" forward while personality (identities now in the foreground) can become an unoticed background. In the next chapters we will explore the major False Core Driver-False Self Compensators and how to begin to dismantle these powerful false selves and liberate our fixated awareness.

CHAPTER IV
QUANTUM PSYCHOLOGY AT A FALSE CORE- FALSE SELF WORKSHOP

The following is an edited transcript, the purpose of which is to present in a workshop format the way that Quantum Psychology interfaces with the False Core. In this way, the reader might be able to experience what it was like to be inside the training room.

MARCH 27, 1994

Now we are going to look at the interface between Quantum Psychology and the False Core. Earlier I spoke about the space inside your body. To begin with, it is important in developing multi-dimensional awareness to split your attention into three or even four ways. One is noticing the external dimension; another part of your awareness can notice your own internal thinking and/or emotional dimensions; and another part of your attention can notice the body while another part can notice where inside your body you feel the **SPACIOUSNESS** of **ESSENCE** or **I AM**. This **SPACIOUSNESS** once (mis)labeled as emptiness is often experienced as a gnawing or painful feeling of **ESSENCE** or **I AM** because of the (mis)label placed on it. Actually the emptiness without a label is the essential quality of **ESSENCE** or **SPACIOUSNESS** and it is also the gateway to the **I AM**, the underlying unity, and Beyond.

I have found that many people have enormous resistance to the **SPACIOUSNESS** which is now (mis)labeled as emptiness inside their physical body because it is mis-perceived and (mis)labeled. This mis-perceived, (mis)labeled inner **SPACIOUSNESS** is now emptiness, and it can be experienced when you go home, or you are having dinner with someone, or it could be when no one is around, or maybe you are just driving your car. But at some point, you might feel this (mis)perceived, (mis)labeled **SPACIOUSNESS** inside "yourself" and probably not like it because of how you label it. Then to try to fix it or fill it up, you might go for food, make a phone call, have sex, or watch TV, etc.—anything to make it go away.

So notice the following: 1) Where in your body do you feel what you probably label as a gnawing inner emptiness? 2) What do you do so you do not have to feel it? 3) What do you do to resist the (mis)perceived, (mis)labeled inner emptiness? Please note, as was mentioned earlier, the (mis)labeled inner emptiness is oftentimes labeled as a lack, derogatory, pejorative, and hence it is resisted. It is not emptiness as emptiness (which is the pure essential quality of **SPACIOUSNESS**). 4) How do you (mis)label this inner **SPACIOUSNESS** as emptiness? 5) How do you attempt to compensate for this (mis)labeled **SPACIOUSNESS** now emptiness (as in a lack)?

So let's go around the room noticing these things:

Wolinsky: (To a student) What do you do to resist the (mis)labeled inner **SPACIOUSNESS**?

Student: Make phone calls; it's become a family joke. Go to the drugstore. The kids may not need something but they might tomorrow. I will run out and do that, so I engage in some type of activity.

Wolinsky: (To another student) You?

Student: I cannot stop doing.

Wolinsky: (To another student) What do you do when you feel the inner emptiness?

Chapter IV

Student: I constantly think, obsess and start creating a problem to think about. Sometimes I think about how to heal this problem.

Wolinsky: You over-think. (To another student) Where in your body do you feel the (mis)perceived emptiness, and what do you do to resist it?

Student: I read. I like to read but there is a compulsive quality to it. I can't stop myself from accumulating information and trying to figure out what others are doing or thinking. Sometimes I get into typing them so I feel more in control and safe.

Wolinsky: (To the group) Notice if you find yourself going into fantasies. Notice what fantasies you might go into in order to resist the (mis)perceived emptiness. The basic issue as we go around the room is, How do you fixate your attention so you don't have to notice the (mis)perceived emptiness. In other words, when emptiness comes up, what do you do to try to *fix it?* Do you go to a future fantasy? What trances do you use to prevent yourself from experiencing the (mis)labeled **SPACIOUSNESS** as emptiness?

Student: It's my eyes. I panic, and I feel it in my eyes, and I start scanning for something to eat. Food.

Wolinsky: Where in your body do you feel it before it goes into your eyes?

Student: I guess I feel it all over my body. It's a total body thing. But mostly in my eyes.

Wolinsky. Okay.

Student: It happens in my breast and there are other varieties. Sometimes I over-think or particularly over-read.

Another student: It depends on what time of day it is that I feel it. But I eat and I exercise, I read.

Another student: My body feels weak in my chest when I feel the emptiness.

Wolinsky: You feel weak. Do you ever try to overcompensate feeling weak by trying to be over-strong?

Woman: Yes.

Another student: I will cook. I'm not a big eater but I will cook.

Another student: I feel the emptiness in my chest, throat, mouth. I become very introverted and I will either think compulsively, read compulsively, or seek sex.

Another student: Gut, eyes, chest, and across the shoulders. I feel the emptiness and then I can't stand it, so I come up with lots and lots of thoughts and images. I usually wind up doing some type of activity, it could be eating, video games, distractions!

Another student: I feel the emptiness in my stomach, eyes, and chest and sometimes my throat. I eat, I get real busy. Thinking, obsessing, I don't read anymore because it's hard to retain it. I used to love to read.

Another student: I feel the emptiness in my chest, and I feel it in my head. I try to fix it or fill it up by creating projects for myself. Developing this workshop or that workshop. Pretty productive stuff I do. And

	also I just read and read and read. Sometimes I watch TV for hours. Nintendo games.
Another student:	When I feel the emptiness, I go inside myself and get real introverted. I think I turn it into depression. But when I am tired of making projects, I may get very sad.
Another student:	I feel the inner emptiness in my upper chest. I then start doing stuff and thinking. Last year or so, playing on the computer replaced having a little car in the back yard to mess with. I play on the computer. It's wonderful. It temporarily makes the emptiness go away.
Another student:	When I feel the emptiness, I bitch a lot like there's something wrong and I want to make it right.
Another student:	When I feel the emptiness, I go through my scheduling book. To see if there are people I can schedule the next day. I try to stay busy doing things.
Another student:	I feel the emptiness in my chest and stomach. I've done all of the above at one time or another. Especially sleeping. Man, I like that sleeping. What I have caught myself doing most recently though is when I experience the emptiness, I label it weak. Then I organize myself and see what am I going to do about it. Sometimes getting to the point where I can sit with it and sometimes not. When I don't, I start making up psychosomatic shit in my head.
Wolinsky:	Psychosomatic shit? Like what?
Student:	Body problems. I've got this, I've got cancer.

Therapeutic Note

I did not pursue this issue in the context of this interaction; however, psychosomatic symptoms are often resistances to underlying emotional states. In other words, when you repress emotional states, you drive the emotions deeper and often into psychosomatic illness. For example, I began one day to feel like I had a flu, nausea, etc. What I discovered was that the underlying state I was resisting was fear. When I experienced the fear as energy, the nausea went away. In Quantum Psychology, we call changing a desire or emotion from the one you have to a more socially acceptable one, a substitution and, if we deny we even have a desire or feeling, it is repression. In homeopathic medicine, when a symptom is suppressed (in my case fear), it goes deeper into the body causing other, deeper reactions, such as my own suppressed fear becoming a feeling of nausea.

Another student: I feel the emptiness in my chest. When I do, I get real busy doing things. Whatever there is to do, I do it or I get out of the house. It's real important to get out of the house.

Another student: Shopping.

Wolinsky: I want to go around the room again one more time. In a word or two, after you label the **SPACIOUSNESS** as emptiness, how do you label the emptiness? Do you label it as, "There is something wrong with me?" "Inadequacy?" Do you label it as some kind of deficit or lack? How do you label the (mis)perceived (mis)labeled emptiness?

Student: Being out of breath, scared.

Student: Out of control.

Student: Unworthy. Unlovable.

Student: Not capable. Not good enough.

Chapter IV

Student: I'm not deserving enough until I know.

Student: Panic.

Student: Depression, inadequate.

Student: Oh my god! And then panic, then loneliness.

Student: Anxiety. I've got to get out of here and be with someone.

Student: What's wrong with me? I compulsively read. Compulsive reading for me is different from reading. When I am compulsively reading, I will read cereal boxes, whatever. Billboards, any words. The secret of the universe is somewhere, it's tacked up on the bulletin board.

Student: I'm damaged in some way.

Student: Like I've been totally rejected by the world and don't exist.

Student: Like I have to read to find out and feel peaceful.

Wolinsky: Looking for peace.

Therapeutic Note

Quantum Psychology calls peace, space, freedom, love, etc., "essential qualities." They are essential because they are aspects of **ESSENCE**. These aspects of **ESSENCE** (see *The Tao of Chaos* and Volume III) are very experienceable once identities are taken apart and reabsorbed into **ESSENCE**. Ichazo calls these qualities "holy ideas." In Quantum Psychology, they are much more than ideas. They are experienceable qualities of your **ESSENCE**.

Student:	I don't know and I really want to know. When I was a little girl, I always wanted to die just for a few minutes so I could go ask God questions.
Student:	The first word that came to mind is scared. I feel trapped—so I don't stand still—keep moving and doing.
Student:	I feel unwanted and unloved.
Student:	I feel anxious. But if I stopped to know what I was feeling, I would feel unwanted.
Student:	Powerless.
Wolinsky:	Powerless. You don't look happy with that.
Student:	No, I'm not thrilled.
Student:	When I feel that feeling, it feels familiar like it's been there a long time. Almost a haunting kind of thing.
Student:	Just kind of low-level depression. It connects with the sadness, it's very familiar.
Student:	Needy and whiny.
Student:	Feeling not enough, then quickly trying to protect myself by getting into doing something which at least shows that I am making an effort.
Wolinsky:	You try to look like you're doing and making an effort and you pretend to do and make an effort?
Student:	Yes, I pretend to be busy, working, doing something.

Chapter IV

Wolinsky: It's false effort?

Therapeutic Note

Notice how all False Self compensating Identities are false because you are overcompensating and *acting*. This is what I call the *Imposter Complex*. This occurs developmentally during the practicing phase when the child has to perform to get love or approval. The problem is that the real state of the child is not acknowledged, but rather the act or performance. Inside the child or adult is the gnawing feeling that they know they are acting or pretending (which they are). Hence, they always feel like an imposter. This, like all False Self overcompensating identities, is false. Like False Selfs, it gives a *false* sense of worth or *false* love to overcompensate for feeling loveless. They are false because they overcompensate for labels placed on the (mis)labeled **SPACIOUSNESS** of **ESSENCE** yielding emptiness (as in a lack), which is not the essential quality of **ESSENCE (SPACIOUSNESS)** or the essence of **ESSENCE** which is the non-verbal **I AM**.

QUANTUM PSYCHOLOGY PRINCIPLE:

Anytime you ignore your underlying state and try to overcompensate through performance or acting, you always know it and feel like a fraud or an imposter.

QUANTUM PSYCHOLOGY PRINCIPLE:

Any attempt to overcompensate or overcome anything through pretending, performance or a False Self-identity, reinforces the underlying resisted state of the False Core and makes you feel worse.

I have met many therapists, engineers, lawyers and doctors, etc., who complain of the *imposter complex*. They need to realize that the *act of doing* feels false to them because it is denying an internal state. Children often only receive love when they achieve something. For the infant, this lack of love is painful. To overcompensate for the pain and alienation or lovelessness, they overachieve. But the love given to them for their achievements feels fake, and even the achievements feel false and not enough. Furthermore, even if appreciation or approval is given for the performance, the person knows it is an act because the act denies and is an attempt to compensate for the internal subjective experience of lovelessness or alienation. Therefore, the praise is not "taken in" because it is an act and a compensation for a painful state. This is why the person is left feeling, "If they only knew what I was really like, they wouldn't think I was so great," or even worse, feeling bad about themselves because they feel like a liar or an imposter. This is why the idea that "we'll fake it till we make it" is disastrous—it only reinforces the imposter complex.

QUANTUM PSYCHOLOGY PRINCIPLE:

The earliest label (False Core) drives the compensating action.

Therapeutic Note

This is why you can never overcome or overcompensate for anything. Why not? Because you put a false label or conclusion there to begin with, and now you are trying to overcome your own false conclusion. In other words, *you are resisting your own false label or conclusion.*

Wolinsky:	Pretending you are doing and making an effort. Not actually subjectively doing or making an effort.
Student:	Yeah, as I find that, it's so clear that with my dad, it was never okay not to be working, producing.
Wolinsky	The doing was false in that you had to look like you were busy.
Student:	Right. You've got to always act busy.

Wolinsky: Act busy. Okay.

Student: Greedy.

Student: Lonely.

Wolinsky: (To the group) As you get further into Quantum Psychology, you have the **VOID** or the **BIG EMPTINESS**. The **VOID** condenses down and gives you an individual, personal, internal **SPACIOUSNESS** prior to the (mis)label of emptiness—which is your personal **ESSENCE** or **I AM**.

Student: It's your God-given **VOID**.

Another student: It's your own little **VOID** inside yourself?

WOLINSKY TO GROUP:

Everything is made of **EMPTINESS** and form is condensed **EMPTINESS**. So this chair is condensed **EMPTINESS**. Our thoughts are condensed **EMPTINESS**. Who you think you are is made of the same condensed **EMPTINESS**. If you could have that experience right now, things would get very spacious. **EMPTINESS** will condense down and make a chair, a thought, a person, a carpet, anger, sadness, and then eventually it will thin-out or dissolve and it will become **EMPTINESS** again. At some point, the **EMPTINESS** might condense down and become a microphone or a Coke; it doesn't really matter. At some point, it will thin out or dissolve and become **EMPTINESS** or the **VOID OF UNDIFFERENTIATED CONSCIOUSNESS** again. That's the nature of **EMPTINESS** or the **VOID OF UNDIFFERENTIATED CONSCIOUSNESS**.

So the question is how do we get in touch with the **BIG VOID**, since it seems so far away. Actually, it's not far away; the (mis)perceived internal emptiness that you feel inside when de-labeled becomes a pathway. If you take off the labels of lonely, inadequate, depressed, unlovable, worthless, off of the (mis)perceived emptiness then underneath that is a "personal" **SPACIOUSNESS** which is **ESSENCE** or **I AM**. *Prior* to you putting your own (mis)label on the **SPACIOUSNESS** like it's bad, unwanted, terrible, weak, sad, empty or whatever. The body develops and organizes around the **I AM** and **ESSENCE**. In that process you lose the awareness of **I AM** or **ESSENCE**. This is a great trauma, the loss of **ESSENCE** and **I AM**.

What happens next? You (the nervous system) create a dissociative observer and corresponding identities whose attention is focused outward, trying to survive. The trauma is the *loss* of **I AM** and **ESSENCE**. Look into your own experiences; most of us feel a sense of loss, like something is wrong, maybe something is missing from us. We believe something must have happened for us to have this feeling. Actually, the *loss* or wrongness feeling is the realization of separation and the *loss* of your **ESSENCE** and **I AM**, the **SPACIOUSNESS** now (mis)labeled as empty. Basically, you have been resisting yourself, by putting a label on **ESSENCE** or **I AM**, like alone, inadequate, loveless, etc. This is the bad feeling we are trying to get rid of. Then we look for a story or a reason or explanation as to what must have happened to us. *We wrongly imagine there has to be a reason why we feel so bad or uncomfortable. The basic feeling is the loss of **ESSENCE** and **I AM**.*

The Quantum Psychology process is one of undoing rather than doing. Identities are formed and in each you get a different observer which is why I call it an observer-identity. *But there isn't one observer, there are multiple observers*. This is one of the major issues that makes Quantum Psychology so different—we see a different observer for each experience. Or as Nisargadatta Maharaj said, "The experienc*ER* is contained within the experience itself." These observer-identities are together, and they are facing "outward" toward the physical world. For example, in order to get love from your mother, you probably have to behave in a certain way. In order to get your father to give you an allowance, you probably had to sit on his lap or laugh at his jokes and tell him how wonderful he was. It was something you had to do in order for the body and individual psychology to survive.

Chapter IV

Quantum Psychology is pragmatic, it focuses on "how to" so that you can learn to work with your False Core-False Self and their accompanying identities.

We are going to go over the way you fixate your attention to a-void the **SPACIOUSNESS** of your own **ESSENCE** and **I AM**. It is how the False Self resists itself by resisting the False Core and its resistance to the **SPACIOUSNESS** now labeled as empty (as in a lack). I'm going to go through them and then you can take a look to see how they fit you.

Rather than trying to figure out what your False Core is, see if you can notice what you do with your attention.[1]

Notice when you feel emptiness, how you label it, and then what you do to overcompensate for the label so you won't have to experience the bad feeling. The process goes like this: You feel the **SPACIOUSNESS**, you label it emptiness, then you label it in some way, usually that it's something bad or unwanted, and then you over-compensate so you don't have to feel the label that you originally placed on the **SPACIOUSNESS**. For example, a False Core-observer labels the **SPACIOUSNESS** as emptiness, then as inadequate, and subsequently tries to prove adequacy by becoming over-intellectual. Another False Core-observer (mis)labels the **SPACIOUSNESS** as emptiness and then as "I don't exist" or "I have nothing" or "I am nothing." Hence, they greedily compensate for having nothing by accumulating thoughts to *prove their existence*. Another False Core-observer might (mis)label the **SPACIOUSNESS** as emptiness, then as "I am alone," and to overcompensate, they over-connect. Another False Core-observer pretends to do. Now, you've done that kind of pretend doing, like cleaning the house. You're not really doing it, you are pretending while a part of you is watching you do it as if through another's eyes. Another False Core-Observer (mis)labels the **SPACIOUSNESS** as emptiness and then (mis)labels it as "I am in-complete" and tries to overcome it by acting overly wise, whole, or trying to get complete, or fill in what is missing.

Once again, the key point is that, whether you are an over-achiever or trying to prove worth, it is all false—false worth, false

---

[1] In Volume III you will see that the False Core-False Self and the observer are not you. And then, they will fall away like a discarded garment of clothing. Then "you" (as pure awareness) are beyond.

actions, false intellectuality, false feeling and false observation etc. Why? Because all of these are defensive dissociative reactions which in Quantum Psychology is called the False Self. All of the False Core-False Selfs are resistances to the (mis)perceived **SPACIOUSNESS** as emptiness, so what you want to do is slow down your process. First, notice where you feel the perceived emptiness in your body. Next, how do you label it? Then what do you do with your attention to a-void the label?

THE THREE-STEP PROCESS

Step I: Notice where you feel the **SPACIOUSNESS** in your body.

Step II: Notice how you label the perceived **SPACIOUSNESS** as emptiness.

Step III: Notice what labels (False Core) you place on the emptiness (as in a lack).

Step IV: Notice what you do with your attention to a-void the label or conclusion placed on the (mis)labeled **SPACIOUSNESS**.

Wolinsky: (To group) There are many qualities of **ESSENCE**. I see them more as actual essential experiences which identities were created to try (unsuccessfully) to get from the outside world. In order to complete this process and stabilize at the level of **ESSENCE**, Quantum Psychology suggests dismantling and reabsorbing the False Core-False Self and their accompanying identities so that you can literally fall into the **NOT-I-I**, the **VOID** and Beyond. It could be seen as: First you dismantle the False Core-False Self and the identities, **ESSENCE, I AM,** and **NOT-I-I** are revealed (see Volume III). Second, reabsorb them into **ESSENCE**. Third, experience the essential quality of **ESSENCE**. These three steps will hope-

Chapter IV

fully help to stabilize this level of your human nature so you can then move on, become aware of and stabilized in the **NOT-I-I** (awareness of the **VOID**). Please note, however, there are no guarantees.

Student: Can you say each one of the identities is seeking to get the qualities of **ESSENCE** but cannot because of their nature and function and so they fall short?

Wolinsky: That's it. Each identity is seeking. It should be noted that seeking or searching is actually part of the nervous system's survival mechanism. Later it can get spiritualized. However, it is survival; for example, to be immortal, live forever, etc. Because it labeled **ESSENCE** a particular way. For example, if it labeled the **SPACIOUSNESS** as emptiness and as imperfect, the identities would search for perfection in the outer world or try to perfect themselves. A loveless False Core and identities seek love. Each identity is seeking essential qualities and **ESSENCE** and, at the same time resisting **ESSENCE**'s label. Identities cannot reach **ESSENCE**. That is one of the reasons for so much pain.

Student: Are we always in one False Core-False Self

Wolinsky: No. Actually you spend lots of times in other False Cores-False Selfs or distractions to the False Core (to be discussed later).

OVERVIEWING: STEPS

1. Notice the Essential Quality of **SPACIOUSNESS**.
2. Notice how it was labeled empty (as in a lack)
3. Notice the False Core-False Self and the identities that were created to defend and attempt to organize against the affixed label and the shock of the Realization of Separation.
4. Dismantle these identities.
5. Reabsorb the identities and the observer back into **ESSENCE**.[2]
6. Experience the essential qualities of your **ESSENCE**.

---

[2] The reabsorption process is described in detail in *The Tao of Chaos*.

CHAPTER V
FUNDAMENTAL UNDERSTANDINGS REVISITED

THE CLIFF NOTES OF THE FALSE CORE-FALSE SELF DILEMMA

Before we go on through the different False Core Drivers and False Self Compensators, let us review the major underlying principles of the process so that our direction is clear and precise.

THE FALSE CORE DRIVER

The False Core Driver is the underlying conclusion, premise, concept, belief, or idea you hold about yourself which drives *all* of your psychology and all you think you are. By being brought to the light of consciousness and "on screen," we can learn to question, enquire, dismantle, and go beyond it. As Nisargadetta Maharaj said, "In order to let go of something, you must first know what it is." The False Core Driver can be likened to the unconscious mind.

The False Core Driver runs on its own, like a machine having been turned on and now on automatic. Quantum Psychology hopefully will help: 1) to stop the machinery called the False Core Drver; 2) enquire into its validity; 3) dismantle it; and 4) go beyond it.

THE FALSE SELF COMPENSATOR

The False Self Compensator is who you think you are or imagine or wish to be. They provide the justification for substituting and compensating behavior, which is actually a defensive persona or mask that hides the False Core Driver. The False Self Compensator can be likened to the conscious mind. It attempts to heal, transform, psychologize or spiritualize itself in an attempt to overcome the False Core. Unfortunately, the False Core-False Self is holographic and the False Self's attempt at healing and transformation only re-enforces it. Why? First, because the False Self's attempt at transformation or healing the False Core is based on believing that the False Core conclusion is true. Second, the False Self is a False Solution to a False Conclusion and, therefore, can yield little, if any, results. Third, because the False Self arose from, and is a reaction to, the False Core—it is part of the False Core. Therefore, any attempt by the False Self to overcome the False Core only re-enforces the False reality of the False Core's conclusions.

QUANTUM PSYCHOLOGY PRINCIPLE:

Any premise (the False Self) which follows a False premise (False Core Conclusion) must also be False.

QUANTUM PSYCHOLOGY PRINCIPLE:

Any attempt by the False Self (the later premise) to heal or transform a False premise (False Core) leads to temporary, unpredictable and unstable results.

For example, if I decide that the only reason why I am unhappy is because I do not have enough money; that is a False conclusion. I then decide that to overcome this False Conclusion (which I do not question but believe) I must work hard and make money which is a False solution to a False Conclusion. This is why money does not make you necessarily happier. *LOOK AROUND!!*

In the same way, if I believe I am separate and suffering *because*, "I am BAD," that is a False Conclusion. If I then decide that if

"I am good," then I will not be separate and I will be in bliss, that is a solution based on a false conclusion. I can then adapt any number of spiritual or psychological process to overcome "I am BAD." However, notice they do not always work. Why? Because the *solution is based on a False conclusion about what the problem is.*

QUANTUM PSYCHOLOGY PRINCIPLE:
The solution to the False Core dilemma is to see the False Core as a False conclusion and a *concept*, an *idea*, which is *NOT YOU*.

Here lies the dilemma: the False Self believes that the False Conclusion of the False Core is true. The False Self "decides" on a "healing," transforming, spiritual, psychological process or strategy, which is based on a False Premise (the False Core).

THE FALSE CORE DISTRACTOR
The False Core Distractor can best be described as a technique and/or feeling that you would prefer to feel so as not to feel the False Core. For example, you might prefer to feel rage rather than the False Core of powerlessness. You might prefer to feel stupid or depressed rather than inadequate. (This process will be clarified later with exercises.)

There are many False Cores. The False Core is where most of your attention is knowingly or unknowingly fixated. The False Self is a compensation, an attempt to heal or transform but ultimately hide the False Core. The False Self is a subtle, seductive, skillful and insidious defense against the shock of the Realization of Separation and the accompanying narcissistic wound by which the False Core began its solidification process.

In the chapters to come the major False Core-False Selfs are described with some therapeutic cases to explore the "how to" go beyond them.

WORKING WITH THE FALSE CORE

Chapter IV took us through a workshop on how these identities were formed through the shock of the Realization of Separation and resistance to **ESSENCE**.

The following chapters have four purposes:

1. To look at each False Core Driver-False Self Compensator.
2. To provide a False Core Driver-False Self Compensator dismantling protocol.
3. To provide a case transcript to see how the dismantling process works.
4. In this way the dismantling process can be used on your own, with another, or within a group context.

It is important to note that the False Core-False Self is a map or a way to define identities which you are not but think you are. Remember the immortal words of Korzybski: "The map is not the territory. . . . The idea is not the thing it is referring to." One must understand that the False Core-False Self *at this level* is a by-product of the nervous system. *It is a description of what is not—not what is*; furthermore, no two False Core-False Selfs are exactly alike just as no two nervous systems are identical. They're like snowflakes and thus the False Core-False Self "shorthand" way to describe how an individual *might* organize his or her world.

It is important to remember that *you are not your False Core Driver or your False Core compensator*. Do not try to improve or make a better False Core conclusion or False Self mask, they are not you, and only perpetuate and prolong your pain. Why? Because 1) if you try to change or improve the False Core, it is the False Self doing it; and 2) if you are trying to make it better or healthier, you subtly must believe it is you. Believing in your False Core-False Self robs you of the ability to "see," "experience," "know," and "appreciate" individual differences. For example, one individual's "I am inadequate" False Core is quite different from another's "I am inadequate." To assume that "they are all the same" is the generalizing aspect of the survival mechanism of the nervous system which robs the individual of his or her own unique nervous system.

Chapter V

To best understand this generalization tendency of the nervous system, Korsbyski created what he called "structural differential"[1] which is a picture of how the nervous system organizes the world. Below is a *very* brief summary:

1. The Quantum unseen level.
   ↓
2. The microscopic level (seen only with instruments).
   ↓
3. The sensation level (non-verbal).
   ↓
4. The object level (seen by you, the nervous system, like a sensation).
   ↓
5. The descriptive level (this is a chair). This is a sensation.
   ↓
6. The inferential level (I) (this sensation is fear, anger, love, etc.)
   ↓
7. Etc. The next inference (II) (anger, fear, is bad, love is good).
   ↓
8. Etc., etc., etc. Inference (III). We must change the bad (anger, fear) into the (love).
   ↓
9. Etc., etc., etc., etc. (Inference level IV). If I can observe this or use a technique, I can change the bad into good.
   ↓
10. Etc., etc., etc., etc., etc. (We can make a never ending number of inferences about inferences.)

What is critical is that each time the nervous system moves from, let's say, the object level (3) to the inference level (7-8), it selects out certain data, omitting other data and generalizing more. The fact is that the *greater the generalization, the further your ideas about the thing, person or experience* is from *what it really is* and, hence, the "further" it is from the Quantum level.

---

[1] This can best be summarized in the books in the biblography.

"People" are not their False Core-False Self. The False Core-False Self is a *descriptive* inference and "the description is not the thing it is referring to."

The inference too is not them nor the thing it is referring to. It is an inference and very far away from who they are at the Quantum level. The nervous system organizes all that you call "you" in a pattern or structure which the False Core-False Self attempts to describe. As Nisargadetta Maharaj said, "In order to find out WHO YOU ARE, you must find out WHO YOU ARE NOT." Quantum Psychology says, "You are not your False Core Driver, False Core Compensator." Basically, use it to find out who you imagine yourself to be but are not and move beyond it. Then WHO YOU ARE will be revealed.[2]

All too often, we study the map and lose sight of the reality (**ESSENCE, I AM**, etc.). I have seen people either trying to figure out their own or someone else's False Core-False Self or caught up in the words used to describe them. All their time is spent discussing the False Self's many attributes not realizing this clever distraction of the False Self to defend against the False Core. Finding your False Core Driver—which can be likened to the organization of the "unconscious mind"—is of utmost importance and it takes time. But while you are determining your False Core, spend time taking apart and dismantling it along with the identities of each type. Why? Because to varying degrees we use all of the False Core-False Self as a defense. Though there is only *one* False Core Driver, there are many False Self compensator and distractor defensive styles. Once you find your False Core, you still have to process through the different identities of each personality type in order to stabilize in **ESSENCE** or **I AM**. So focus on **ESSENCE** and **I AM**, etc., and do not get caught up in believing the map or in Quantum Psychology.

Problems occur when we get attached to the means, be that a yoga, a psychology, a mantra, a yantra, a tantra, or a teacher. When we develop these kinds of attachments, we wind up worshiping the method and lose sight of the goal.

With this in mind, Quantum Psychology is *short* on descriptions of each False Core-False Self and *long* on how to process your-

---

[2]As will be discussed in Volume III, even the user, noticer and inquirer of the False Core-False Self is not you and will fall away naturally

Chapter V

self out of your False Core-False Self and identities and stabilize in **ESSENCE, I AM**, etc. Instead of becoming attached to the method, look at how to unhook the False Core-False Self which drains your energy and causes you to lose awareness of other dimensions.

PROTOCOLS FOR DISMANTLING FALSE CORES

Quantum Psychology provides some basic understanding taken from Nisargadatta Maharaj to help to take apart and dismantle the False Core-False Self:

MAJOR UNDERSTANDINGS (FACING OUTWARD):

Quantum Psychology
1. You are not your False Core-False Self.
2. The False Core was formed to explain "why the narcissistic injury occurred.
3. Someone modeled the False Core-False Self for you and you "took on" and fused with someone else's False Core-False Self.
4. There might be an genetic-energetic pre-disposition to a False Core-False Self.
5. There is a lineage of generations of the False Core-False Self. (This is called a miasm in homeopathic terms.)

Nisargadatta Maharaj
1. Anything you think you are, you are not.
2. Whatever you know about yourself came from outside of you; therefore discard it.

Major Contexts
1. Determine who or what was the "model" for the False Core.
2. Make the implicit-explicit.

3. Allow the False Core-False Self to go back to the lineage.
4. Handle the "shock" of the narcissistic injury.
5. The body "holds" the False Core-False Self.
6. Turn your attention around and notice what, if anything, is there.

MAJOR UNDERSTANDINGS (FACING INWARD):

"To go beyond the mind, one must look away from the mind and its contents."
　　　　　　　　　　　　　　　　Nisargatta Maharaj

1. **ESSENCE** or the (mis)labeled **SPACIOUSNESS** now labeled emptiness and/or **I AM** are blamed for the "shock" (realization of separation).
2. The False Self-compensating identities face outward to A) defend against the shock and; B) to try to handle (heal) the False Core Driver.
3. The (mis)labeled spacioiusness becomes through the label emptiness (as in a lack). In this way, **ESSENCE** and **I AM** are labeled the same as the False Core (i.e., the reason for the realization of separation) and is A-VOIDed, i.e., covers the awareness of the **VOID**, and is resisted.

Major Approaches
1. Notice how **ESSENCE** and **I AM** are blamed.
2. Notice the labels placed on **ESSENCE** and how those labels prevent the emergence of **ESSENCE**.
3. Peel back and face the (mis)labeled **SPACIOUSNESS** which is now the emptiness of **ESSENCE**. The (mis)perceived emptiness is inaccurately called the existential crisis because it is not experienced as **SPACIOUSNESS** but as the label placed on **SPACIOUSNESS** which is now emp-

tiness (i.e., depression, alone, etc.). It is only a crisis if you do not go beyond the (mis)label placed on the **SPACIOUSNESS** of **ESSENCE** which is actually the *fullness* of Essential Qualities because it is not understood that the emptiness with no label is **SPACIOUSNESS**, and it provides a pathway to go beyond "yourself" (i.e., False Core-False Self and, paradoxically, an entry point to the liberation of awareness).
4. Experience **ESSENCE** as possessing the qualities that the False Core-False Self is seeking. Einstein stated, "You cannot solve a problem at the same level of consciousness which created it." **ESSENCE** is the cure for the shock of the Realization of Separation because from **ESSENCE** you can have an **ESSENCE** to **ESSENCE** unity with mom, while simultaneously being separate at a biological, thinking and emotional level.
5. Re-absorption of False Core-False Self identities, *back into* **ESSENCE**.
6. Going beyond **ESSENCE** into the Essence of **ESSENCE,** the **I AM**, i.e., with no thoughts, memory, emotions, associations or perceptions.
7. The development of multi-dimension awareness.

SPLITTING

The False Core Driver (unconscious mind) is split and oftentimes remains separated from the False Self (conscious mind) by a layer of amnesia or blankness. Quantum Psychology is not into the unconscious mind. We see the unconscious as all you do not want to know and experience which is being driven by the False Core Driver and the shock of the Realization of Separation. The False Core is split-off and surrounded by a layer of amnesia to protect it from the realization of separation. On the other side of the amnesia lies the False Self Compensator. This can best be illustrated below.

Splitting is a phenomenon which occurs when you are in one identity and forget or don't know that the other identity exists. It's like the old saying, "The left hand doesn't know what the right hand is doing." In this case, the compensating identity oftentimes doesn't know or want to know that the False Core exists and, hence, has an *insight barrier*.

There is a major splitting-off which occurs during the shock of realization of separation when the False Core is solidified from its formless energetic state and **ESSENCE** is blamed. There is a split between the False Core and the False Self Compensator. Quantum Psychology hopefully helps us to see the split as well as the relationship between the two, and to go beyond.

Splitting happens with all identities but it is most pronounced with the False Core and False Self. To illustrate this, in 1975 several of my Arica friends met Swami Muktananda. Their impression was, "I thought he was fabulous, very insightful, but he was *in* puritan." Which was accurate. He was in a puritan identity. Later his hedonist identity came out but he was unable to acknowledge this. And so, while being in the puritan identity, he attempted to handle the hedonist identity through *spiritualization*. He said that sex was tantra, and projected his hedonist identity on others, accusing them of being "impure" for even thinking such thoughts of him. He thus projected his hedonist identity on others.

In less dramatic terms, if you are *in* the "I love myself identity," the "I hate myself identity" does not exist. In fact, when you are *in* "I love myself," You cannot conceive of the fact that you hated yourself. Three days later when you are *in* "I hate myself" then you cannot conceive of the fact that "I love myself" even existed. This is the splitting phenomena where *between each identity is a layer of amnesia*. When you are in one, the other doesn't exist. It's like two separate universes.

Chapter V

I once worked with a woman who complained about money. One minute she would say, "I'm not into money." The next minute, "I am not into money—you are into money." When I told her that a few days ago she was arguing with me about wanting more money, she got confused and upset with me. She was amnesic about her earlier remark. In short, there was a splitting and an amnesia between the two Identities.

Projecting a disowned "part" is called Projective Identification. In the earlier examples, Jimmy Swaggart cried out that the devil made him do it. *It was the devil (really himself). He was a victim of something he disowned which now seemed outside him and outside of his control and which he now attributed to and spiritualized as the devil.*

TRACING YOUR FALSE CORE

Processing your False Core (driver) and its accompanying False Self (compensator) is imperative.

Tracing your False Core requires drawing a line and making a connection between the False Core and False Self Compensator. This normally cannot be drawn or seen because of amnesia which prevents you from knowing about how they are related to one another.

Since the False Core is so defended, all unwanted experiences are relegated to it. An amnesic blankness prevents one from seeing the inter-connection of them. Things remain as one holographic unit unknowingly fused together by the associational trance until the trance is broken.

This amnesia can be experienced in any or all of the trances mentioned in *Trances People Live*, or even the dissociated void where amnesia and blankness are *spiritualized*. This amnesia when gone through and beyond, through tracing the False Self to the False Core, allows the entire holographic unit to be seen, realized and, ultimately, gone beyond. When this is not known or viewed, the connection between the False Core Driver and False Self Compensator are forever split off from each other, resulting in the inability to see the two sides of the same coin.

We can suggest as a *theory or story* that we are born in utero with a genetic-energetic predisposition to our False Core. I do not want to suggest that we *always* "take on" our parents' False Cores; but there are often similarities in movement which mirror them. To locate what your False Core might be, notice the experience that you are having and ask yourself, *"What's the worst of it?"* If you continue to pursue this enquiry, you will eventually get back to the False Core. It's a way of tracing behavior, thought, emotions, fantasies, etc., which are driven by your False Core, and which will lead you back to it. Continue to ask yourself, "What's the worst of it?" Internal states which we normally label and then experience as unpleasant can motivate us to find out what our False Core Driver is.

> As the Tantric Yoga of Kashmier stated, "At moments of extreme anger, joy, fear, or when you are running for your life, if at that moment you become introverted, you will experience Spanda, the divine throb." (Spanda Karikas, p. 5)

If you stay with a painful experience and keep asking "What's the worst of it?" or "What is so bad about that?" until you reach bottom, you will reach your False Core Driver. As you practice more, you will learn to circumvent everything, tracing it all back to the False Core. When you arrive in the False Core, observe it but do not believe it. Be it, un-be it and then stay in the nonverbal experience just prior to it, the **NON-VERBAL I AM** (no thoughts, memory, emotions, associations or perceptions (see Volume III)). This will put a halt to the automatic abstracting, selecting and omitting process of the nervous system (i.e., moving outward into descriptions, explanations, stories, inferences, etc.) and cut the False Core at the root, just *prior* to the VERBAL **I AM** (i.e., the **NON-VERBAL I AM** or pure beingness, (to be discussed in Volume III)).

TRACING YOUR STRUCTURE

The clearer you are about the False Core Driver, its connection to the False Self Compensator as well as the False Core defenses

Chapter V

and distractions, the easier it will be to be to trace your entire psychology back to this one Organizing Principle or Structure. When you trace back to that one structure by letting go of all associations, its strength begins diminishing and the (mis)labeled **SPACIOUS** now emptiness (as in a lack) dissolves, revealing the (**SPACIOUS**) Essential Qualities of **ESSENCE**, the No-State State of **I AM** and the underlying unity of the **NOT-I-I** which moves forward into your awareness. For most people, the False Core is in the foreground and **ESSENCE** and **I AM** is in the background. When the False Core-False Self is dismantled, this will be reversed and the **ESSENCE** and **I AM** become foreground and the False Core-False Self an unnoticed background.

Always remember that *you are not your False Core-False Self*.

Once you can shift your focus backward *prior* to your stories and concepts and remain in the **NON-VERBAL I AM** background, the False Core unravels and "you" are beyond your "personal" psychology. When "I" was with Nisargadetta Maharaj, he once asked a student, "What is the title of my next book?" "Beyond Consciousness," she said. "No," he said, "*Prior to Consciousness*. Prior to your last thought—stay there."

The Indian sage, Ramana Maharishi, died in 1950. In the early 1980's, I went to his Ashram and while there I read a story about him: There was once a seeker from Europe who came to meet him. After much suffering, this seeker finally arrived at the mountain known as Arunachala where Ramana Maharishi lived. He approached the sage and said, "I have been seeking for years. I want to find out who I am." "Go back the way you came," Maharishi told him, "Just go back the way you came." When his disciples heard what he said, they were upset by his apparent lack of compassion. But in fact Maharishi was saying, "Trace it back, this will lead to **ESSENCE, I AM, NOT-I-I** and, ultimately, to the **VOID OF UNDIFFERENTIATED CONSCIOUSNESS.**"

Getting your False Core is a major breakthrough. It gives you the one structure which organizes your subjective internal experiences. To keep on tracing it back requires inner focus. When you are able to do this and remain in the **VERBAL I AM** and later in the

NON-VERBAL I AM *prior* to the False Core Driver-False Self Compensator, your entire psychology begins to fall away (it has no ground to land on). This practice can be done all the time—lying in bed, walking, sitting, etc. No matter where you are, you can trace every experience back to the False Core structure. The story as to why you feel what you feel is not important. The key is the False Core structure.

TRACING THE FALSE CORE

THE STEPS

Step I: Trace your experience back to the False Core Driver.
Step II: Acknowledge, own, and observe the False Core.
Step III: Disown the False Core. This means "getting" that the False Core is not you and UN-BEing the False Core.

Do not believe the story line. The story line is only the reason (see Volume III), justification and seductive defensive strategy which is part of the False Self Compensator and False Core distractors. *The story-line comes, neurologically speaking (in neurological time), after the experience has already past the moment of space-time that it occured in.* This is why the first exercise in Quantum Psychology training is to take your attention off the story as to why you feel what you feel and put your attention on the feeling itself. This is done so that you can learn to develop *a functional awareness*, i.e., where you put your attention. If you have fixated your awareness on the story and believe the story and the False Self, then the story becomes the way you resist experiencing your False Core. But actually the story is an abstraction of the nervous system, a justification for what you are experiencing and the story arises *after the the moment in space-time that the experience has occured in.*

If I believe the story, then I am always going to be justifying my position, trying to overcome the False Core. For example, "Maybe my problem is my job. If I get another job, I won't feel so inadequate." "Maybe if I have another relationship, I won't feel so alone."

Chapter V

Here the story lines of jobs or relationships are used to distract you from the False Core which drives the psychology.

QUANTUM PSYCHOLOGY PRINCIPLE:
The storyline distracts you from not only experiences, but the **NON-VERBAL I AM**.

By believing the story (False Core) about why you feel what you feel, you will be kept from feeling what you feel. Then you will try to rectify the situation (False Self) with something outside of yourself.

Step V: Be in the *VERBAL* **I AM** *prior* to the False Core: I am *(fill in the blank)*.
Step VI: Go into the **NON-VERBAL I AM** *prior* to the verbal **I AM** and the False Core (i.e., no thoughts, memory, emotions, assocations or perceptions) (see Volume III).

QUANTUM PSYCHOLOGY PRINCIPLE:
Everything is filtered through the False Core.

When you get into explanations, stories, reasons or justifications of "why," then you are *splitting off* from the experience itself and moving into the "newer brain." My psychology mentor, Dr. Eric Marcus, once said to me "Give up *why* and *because*."

Until it is dismantled you will always "do" your False Core Driver-False Self Compensator. You cannot not do it. The whole set up is to do it. All states and Identities begin with the False Core Driver. You cannot think yourself out of the False Core-False Self because *it* is doing the thinking. Your mind is driven by it. The only way out of *it* is to dismantle it and enter into **ESSENCE** or **I AM** which is beyond the False Core.

EVEN DEEPER

As previously mentioned, to find the False Core, you need to ask yourself "What is the worst of it?" with each experience. You can fine tune this process and go even deeper by asking yourself after each thought, impulse, behavior, emotion, or fantasy, etc., "What assumptions underlie this thought?

Example 1

Client: I am really angry at Bill (a fellow employee). He is so self-centered.

Wolinsky: This idea of self-centered and angry, what's the worst of his being self centered?

Client; That I don't get what I want being around him.

Wolinsky: And what is the worst of not getting what you want?

Client: It is like I am unimportant.

Wolinsky: And what's the worst of being unimportant?

Client; "*I don't exist*" around him.

Wolinsky: Is that the bottom of it?

Client; Yes.

Example 2

Client: I am so hurt that she assumed and made a judgment about who I am.

Wolinsky: And what's the worst of someone making a judgment about who you are?

Chapter V

Client:	That I feel like I am not seen.
Wolinsky:	And what's the worst of not being seen?
Client:	That "I don't exist."
Wolinsky:	And what's the worst of non-existence?
Client;	That "*I am alone.*"
Wolinsky:	Is that it?
Client;	Yes.

CONCLUSION

I hope this chapter has served as a summary of what has been gone over earlier and the underlying principles of "how to" handle what is to come—i.e., the descriptions and the context for the questions which will give us the "how to" go beyond the False Core-False Self.

Good luck.
See you soon
Your brother,
Stephen

THE FALSE CORE DRIVERS-
FALSE SELF COMPENSATORS

CHAPTER VI
THE WAY OF
THE IMPERFECTIONIST

FALSE CORE DRIVER:
"I am imperfect." "There must be something wrong with me."

FALSE SELF COMPENSATOR:
I must prove I am *not* imperfect, and that there is *not* something wrong with me. I must be perfect.

STRATEGY:
Seek internal and/or external perfection.

The tongue-in-cheek use of the word *path* brings us to the obsessive-compulsive nature of the underlying premise of all False Cores. In this case, the False Core Driver "There must be something wrong with me, I am Imperfect," both resists and, desires its own repetition. This False Core Driver, like all the others, bases its entire psychology (thoughts, emotions, associations, body actions, fantasies, dreams, wishes, desires, etc.) on a false premise and conclusion. Since this premise arises and is solidified so early in life and resisted due to its association and fusion with the shock of the Realization of Separation, the False Core of "There must be something wrong with me, I am imperfect" formulates that the way to Nirvana (merging with mom, later if *psychologized* it becomes getting healthy and getting what you want, or if *spiritualized* with God, if *relationship-ized* with another, if *work-ized* with a job) is to be perfect. This solution is imagined to be accomplished through the False Self Compensator trying to prove I am not imperfect, and that there is nothing wrong with me, and "I am perfect."

The problem is that the first formulated premise—in this case "There must be something wrong with me, I am imperfect"—is the strongest and organizes and rules this closed loop. Subjectively speaking, there is no way out. Why? Because this False Core premise always proves itself. In other words, it will always subjectively prove that "I am imperfect, there must be something wrong with me."

Since the shock of the Realization of Separation and its accompanying premise that "I am imperfect, there must be something wrong with me," is resisted, an obsessive compulsive tendency emerges. As with all obsessive-compulsive tendencies, the underlying belief is, "If I can just do it again, maybe I can do it *right*; right the wrong and things will be different." Thus, the False Self Compensator is created to resist the initial shock of the Realization of Separation and overcome the False Core Conclusion. For this reason, the False Core "I am imperfect, there must be something wrong with me" is reenacted, resisted, and recreated as a repetition compulsion—acted out again and again and again in the hope of having another outcome.

To counter this premise, the major formulation of the False Self Compensator is, "I have to be perfect or prove I am not imperfect or defective," or "If only *they* or it were perfect (projected premise of imperfection on another), Nirvana (the merger with mom) would occur."

It can be considered as a *possible* hypothesis that therapies which recreate the trauma and place a "happy or desired" ending on the story (drama) can add insight to the problem, but in the final analysis, it is the False Self who places the happy ending on the story (drama) to overcome the False Core. This only serves to re-enforce the premise by using the False Self Compensator and its obsessive-compulsive tendency which attempts to overcome the False Core Driver. In short, they "act out" the story (drama) as to why "I am imperfect, there must be something wrong with me, I must be perfect" bind, not realizing the story itself was created by the nervous system and comes neurologically after the fact to justify itself. I am not trying to "put down" any form of therapy. I am, however, proposing that if our intention is to find out who we are, then the underlying assumption of any psycho-spiritual system should not be left unexamined.

THE REASONS GIVEN FOR THE FALSE CORE

"There is something wrong with me because (*fill in the blank*). I am damaged, imperfect, empty, I am), etc.

The False Core Driver always follows the Way of the Imperfectionist in order to *prove and recreate itself*. It has a deep feeling of imperfection. The False Self is going to make all kinds of moves to resist, resolve, reinforce and re-enact the False Core in an attempt to heal the narcissistic wound created during the shock of the Realization of Separation. To discover your False Core, notice what it is that really pulls your chain of associations. What is it that is driving you? What am I always trying to correct or not allow, in this situation, the feeling "there is something wrong with me?"

THE PROJECTION

"There must be something wrong with me" can be projected outside of yourself in relationships. For example, "There is nothing wrong with me, there's something wrong with them." The illusion is that if my partner changed and were a perfect (good enough) mom or dad—which trance-lates to merging with me, their being my reflection or matching the image I have of them, or in short, playing my game—then the pain of the shock of the Realization of Separation and the False Core would be healed. In other words, when "There is something wrong with *them*" is projected onto a partner, people imagine that if their imperfections were not there, everything would be perfect.[1]

---

[1] It should be noted that some psychotherapy schools and systems suggest that the therapists should act or be the good enough Mom or Dad. This strategy is a pretence on the part of the therapist because they are doing it for money and it is an outright lie, they are NOT MOM OR DAD. In this way, the client-therapist can easily re-enact the past through their trance-ference-counter-transference issues. For this reason Quantum Psychology does not support this delusion or its outcome of an age-regressed trance-ference by the client or student or an age-regressed counter-tranceference of a therapist, teacher or Guru

FALSE CORE DISTRACTORS

Distractors are the states within each False Core which distract you from the False Core Driver. With this False Core Driver, *resentment* toward mom is generalized to (*fill in the blank*), for not merging and being a perfect reflection. Being perfect means you're my reflection. Simply put, if you are not my reflection, there is something wrong with you or me so I must either get you to change or I must be your perfect reflection to get you to merge. Resentment is used to distract one from this False Core Driver and the experience of "There is something wrong with me." In other words, "There is something wrong with *you*," that is why I am resentful and feel pain, and if I can get you to see and change what I see as wrong with you, then "my" pain will go away.

FALSE CORE DISTRACTOR EMOTION: RESENTMENT

FALSE SELF IDENTITIES

False Self Identities are strategies which compensate and distract you from your False Core.

Oscar Ichazo presents two major dichotomies (identity pairs). The first he calls the "rigid," and the second, the "sentimental."

Quantum Psychology distractors:
The first is the *distancer* and the second is the *merger*.[2]

This False Core is clingy, but its body is rigidified. It is very controlling, trying to perfect its external world by projecting imperfection, or to perfect its interior world by trying to control the exterior world and preventing imperfections (things which do not match their image) from occurring. They are going to appear from the outside as distant, but actually they are inwardly clingy, wanting to be perfect or to perfect others so they can merge and heal.

---

[2] Please note, Quantum Psychology credits Ichazo for the discovery of these distractor-identities. We only changed the labels a little in an attempt to clarify, but the discovery is clearly his.

Chapter VI

Please remember that the merging and healing attempts by the False Self Compensator are an integrated age regression. That through "spirituality" (spiritualization) or "healing," the age regressed adult unknowingly comes up with a socially acceptable way to do the impossible—*merge with Mom*. This *merging delusion, i.e., if I am perfect or perfect myself then I can merge*, is a never ending cycle and can take the form of merging with GOD (spiritualization) or healing the wound (psychotherapy). However, these attempts are integrated age regressions which is why, after years of psychological or spiritual practice, there is usually no real satisfaction. This is because *the work is generated from and done by the False Self Compensator* and its integrated socially acceptable mask is an attempt to overcome the False Core Driver and the shock of the Realization of Separation and to merge with mom.

HEALING STYLE

The healing style is the way you try to heal the narcissistic wound through re-merger, namely, the False Self Compensator "If I can just be perfect," "Do it perfectly," then I will be healed and I will merge with mom/God. This is the unconscious preoccupation and motivation of this particular False Self Compensator.

SPRITUAL PRACTICE: TO BE PERFECT WHICH RE-ENFORCES THE FALSE CORE DRIVER-FALSE SELF COMPENSATOR

THE OVER-PERFECTIONIST

Oscar Ichazo calls this the over-perfectionist, its passion being resentment.

This False Core is the way attention is unconsciously fixated through the shock of Realization of Separation (and the loss of the awareness of **ESSENCE**). Why? Because perfection was "I and mom" (later *spiritualized* as God) were one. Everything after the separation realization is unknowingly compared from the False Core and has to come up short since the "I am one" is now the standard reference point. The unconscious fixation of attention caused by the shock creates a splitting-off and a layer of amensia. Out of that amnesia comes the False Self Compensator of the over-perfectionist. In this way, Ichazo's underlying state of resentment associated with this reaction "blaming the self or others," is that somehow you or another are defective and imperfect and, therefore, fell from grace (being one with mom/God).

ESSENCE is labeled "imperfect" by the observer and this comes from the observer-False Core-False Self-complex. For each False Core, I will present Quantum Psychology's False Core Driver-False Self Compensator, the major distractor identities along with Ichazo's major identities.

Quantum Psychology
False Core Driver—"There must be something wrong with me."
False Self Compensator—Trying to prove perfection by becoming overly-perfectionistic.

Ichazo's Identities
Identity #1—rigid
Identity #2—sentimental

Chapter VI

From this point on we will write Identity as I-dentity to denote it as an "I" which *at this level* is a construction of the nervous system, and which arises in a later and different moment of space-time than the experience. In this way, its construction is used to organize, justify, and give a reason for the unwanted experience.

Once the reason is formulated, (False Core) we can make sure it will never happen again, (False Self). You are prior to this construction and are not the False-Core-False-Self I-dentity.

DISMANTLING THE FALSE CORE-FALSE SELF I-DENTITIES: THE PROTOCOLS

The idea of going beyond polarities is not new. We see it as early as the 12th Century in Patangali's Yoga Psychology (Sutras).

Here are some basic I-dentity questions to explore and unlock the False Core-False Self.

One of the main things Nisargadetta Maharaj taught me was how to enquire. Keeping within this tradition, the following questions should be considered almost mini-meditations. The purpose is to ask yourself a question and notice what, if anything, "pops up" and then "acknowledge it and discard it." One should not analyze or search for an answer. Rather just notice what "pops up." In this way, we can acknowledge what has been unknowingly pulling our attention, and then become free of it. To paraphrase the Indian Sage Meher Baba, the ego (I-dentity) is like an iceberg: 90 percent of it is underwater. Once brought into the light (awareness), the iceberg can begin to melt. Quantum Psychology says that through deliberate enquiry, the ego (False Core-False Self), which has been holding part of our awareness, is freed—thus freeing our awareness and freeing us.

With this in mind, ask yourself the questions, noticing what emerges. Continue with each question until nothing more emerges. Then, move on to the next.

FALSE CORE DRIVER—"THERE MUST BE SOMETHING WRONG WITH ME"

1. Where in your body do you feel the *"There must be something wrong with me"* I-dentity? Scan your body and notice where the I-dentity is.
2. Who modeled the *"There must be something wrong with me"* I-dentity for you? Ask yourself this question and write down the answer until nothing pops up.

Therapeutic Note

At this point it is advantageous to externalize the *I*-dentity. This can be done in the following way:

Step I: Notice where in the present time body the *I*-dentity is located.
Step II: Notice the *I*-dentity's size and shape.
Step III: Take the label off of the *I*-dentity. Have it as energy.
Step IV: Allow the *I*-dentity as energy to move from the present time body to another physical location in the room, and then to become solid again.
Step V: Continue the rest of the enquiry protocols, asking the *I*-dentity the questions and allowing the *I-dentity to respond.*

3. Has the *I*-dentity *"There must be something wrong with me"* ever justified separation from another? Ask the *I*-dentity this question and write down the answer until nothing pops up.

4. Has this *"There must be something wrong with me"* I-dentity created beliefs to try to maintain an individual self? If yes, write down the beliefs. Ask the *I*-dentity this question and write down the answer until nothing pops up.

5. Has the *"There must be something wrong with me"* I-dentity ever resisted feeling? If yes, write down what feelings. Notice if there is a shock. Ask the *I*-dentity this question and write down the answer until nothing pops up.

6. Has the *"There must be something wrong with me"* I-dentity ever created a spiritual or psychological philosophy, path, technique or approach to justify or make more real its existence? Ask the *I*-dentity this question and write down the answer until nothing pops up.

7. Has this *"There must be something wrong with me"* I-dentity ever tried to get even, punish another or tried to control another? Ask the *I*-dentity this question and write down the answer until nothing pops up.

8. Has the *"There must be something wrong with me" I-dentity* ever come-up with concepts, ideas or beliefs as a way of feeling special, more spiritual, or different from another? Ask the *I*-dentity this question and write down the concepts until nothing pops up.

9. Has the *"There must be something wrong with me" I-dentity* ever tried to get you to believe that you were a personality? Ask the *I*-dentity this question and write down the answer until nothing pops up.

10. Prior to you taking on the *"There must be something wrong with me" I-dentity*, was there any chaos that occurred? Write down the answer until nothing pops up.

11. During the taking on of the *"There must be something wrong with me" I-dentity*, what beliefs got created? Write down the answer until nothing pops up.

12. During the taking on of the *"There must be something wrong with me" I-dentity*, what decisions were made? Write down the answer until nothing pops up.

13. Notice that the *"There must be something wrong with me" I-dentity* is a mask, act and/or a False conclusion.

14. Where in your body do you experience the **SPACIOUSNESS**? Write down the answer until nothing pops up.

15. Notice the *"There must be something wrong with me" I-dentity* (mis)labels the **SPACIOUSNESS** as emptiness (as in a lack).

16. In present time, peel back the label of emptiness placed on the **SPACIOUSNESS** and go into the **SPACIOUSNESS** inside the physical body. View the (mis)label of emptiness and/or the *"There must be something wrong with me" I-dentity* which was placed on the **SPACIOUSNESS** from "back there" (inside the **SPACIOUSNESS**).

Chapter VI

17. How does the label and the *"There must be something wrong with me" I-dentity* seem to you from "back there"? Write down the answer until nothing pops up.

18. Ask the *"There must be something wrong with me" I-dentity*, "What are you seeking more than anything else in the world, and allow the I-dentity to answer?" Ask the I-dentity this question and write down the answer until nothing pops up.

19. Ask the emptiness label and the *"There must be something wrong with me"* False Core, "What are you seeking more than anything else in the world?" Ask them this question, allow them to answer and write down the answer until nothing pops up.

20. From inside the **SPACIOUSNESS**, feel the quality which the *"There must be something wrong with me" I-dentity* and the label of emptiness are seeking. Then, notice the size and shape of the False Core emptiness label, and the size and shape of the *"There must be something wrong with me" I-dentity* (False Core); and also notice the **SPACIOUSNESS** that they are floating in.

21. See the *"There must be something wrong with me" I-dentity* and the emptiness label as being made of the same substance as the **SPACIOUSNESS** while you continue to feel the essential quality that the *"There must be something wrong with me" I-dentity* and the label were seeking.

QUANTUM PSYCHOLOGY FALSE SELF COMPENSATOR— "I HAVE TO BE PERFECT" I-DENTITY

1. Where in your body do you feel the *"I have to be perfect" I-dentity*? Scan your body and notice where the I-dentity is.

2. Who modeled this *"I have to be perfect" I-dentity* for you? Ask yourself this question and write down the answer until nothing pops up.

3. Have you ever used this *"I have to be perfect" I-dentity* as a way to prevent people from separating from you? Ask yourself this question and write down the answer until nothing pops up.

4. Has this *"I have to be perfect" I-dentity* ever been used as a way to have a relationship? Ask the *I*-dentity this question and write down the answer.

5. Has this *"I have to be perfect" I-dentity* ever been used as a way to overcome a feeling? If so, write down which ones. Ask the *I*-dentity this question and write down the answer until nothing pops up.

6. Has this *"I have to be perfect" I-dentity* ever been used to avoid feeling separate or to become one? Ask the *I*-dentity this question and write down the answer until nothing pops up.

7. Has this *"I have to be perfect" I-dentity* ever been used to make itself feel important to another person so that the other would not want to be separate? Ask the *I*-dentity this question and write down the answer until nothing pops up.

8. Has this *"I have to be perfect" I-dentity* ever been used as a way of proving its individuality? Ask the *I*-dentity this question and write down the answer until nothing pops up.

9. Has this *"I have to be perfect" I-dentity* ever been used to avoid or prevent or make others afraid to confront it? Ask the *I*-dentity this question and write down the answer until nothing pops up.

10. Has this *"I have to be perfect" I-dentity* ever been used as the best way to merge with another or imagine if I am perfect, oth-

Chapter VI

ers will want to merge with me? Ask the *I*-dentity this question until nothing else pops up.

11. Has this *"I have to be perfect" I-dentity* ever been used to control another? Ask the *I*-dentity this question and write down the answer until nothing pops up.

12. During the process of taking on this *"I have to be perfect" I-dentity* was there any chaos that was occurring (i.e., shock of separation)? Write down what chaos.

13. What decisions or beliefs emerged during the taking on of the *"I have to be perfect" I-dentity.* Ask the I-dentity this question and write down the answer until nothing pops up.

14. By taking on the *"I have to be perfect" I-dentity* what was being avoided? Ask the I-dentity this question and write down the answer until nothing pops up.

15. During the taking on of the *"I have to be perfect" I-dentity* was there any resistance to the (mis)label of emptiness? Ask the I-dentity this question and write down the answer until nothing pops up.

16. Has the *"I have to be perfect" I-dentity* ever been used as a way of feeling spiritual, or as a way to justify a spiritual goal or practice? Ask the *I*-dentity this question and write down the answer until nothing pops up.

17. Notice how the **SPACIOUSNESS** was (mis)labeled.

18. *Intentionally* place the *I*-dentities and labels on top of the **SPACIOUSNESS**.

19. Where in the body do you experience the **SPACIOUSNESS**.

20. From inside the **SPACIOUSNESS** inside your body, how does the (mis)labeled emptiness, the *"I have to be perfect" I-dentity* seem now? Ask the *I*-dentity this question and write down the answer until nothing pops up.

21. Ask the (mis)labeled emptiness and the False Self of *"I have to be perfect" I-dentity*, "What are you seeking more than anything else in the world?" and *allow them to answer*.

22. Be in the essential quality of **SPACIOUSNESS** and experience the quality of **ESSENCE**, the *"I have to be perfect" (False Self) I-dentity* and the emptiness label are seeking.

23. From inside the **SPACIOUSNESS** of **ESSENCE**, notice the shape and size of the *"I have to be perfect" (False Self) I-dentity* and the emptiness label.

24. Notice the **SPACIOUSNESS** surrounding the *"I have to be perfect" I-dentity* and the emptiness label. See the *"I have to be perfect" I-dentity* and the emptiness (mis)label as made of the same substance as the **SPACIOUSNESS** *while you feel the essential quality of* **ESSENCE**.

CONCLUSION

1. Without using your thoughts, memory, emotions, associations or perceptions, is there something wrong with you, nothing wrong with you or neither?

2. Without using your thoughts, memory, emotions, associations or perceptions, are you perfect, imperfect or neither?

3. Without using your thoughts, memory, emotions, associations or perceptions, what does perfection or imperfection even mean?

4. Without using your thoughts, memory, emotions, associations or perceptions, what is emptiness?

Chapter VI

**NOTICE THE
NO-STATE STATE OF
THE NON-VERBAL I AM**
(Volume III).

CHAPTER VII
THE WAY OF THE WORTHLESS

FALSE CORE DRIVER:
I am worthless, I have no value.

FALSE CORE COMPENSATOR:
I must prove I am not worthless and that I have worth and value.

This False Core Driver has the premise of "I am worthless." It describes a conclusion which *theoretically* began to solidify between the age of 5-12 months. "I" use the word "theoretically because it too is only a "story," conclusion or inference and can never actually be proved.

The False Core of "I am worthless" uses the False Self of "I have to prove worth or I am worthy" to prove it has value. If this strategy fails, they will be left with the False Core of "I am worthless." Whatever attempt is made to *overcome* this, whether by becoming an over-giver, by making others dependent on you by acting or by telling yourself or others how great, beautiful, and successful you are or they are, or by trying to create an image of worth (wealth) or value to hide the internal state of worthlessness-no value, or attempts to get others to say you have value (flattery) *I am worthless prevails!!* Why? For two reasons: 1) The first premise (False Core Driver) is the strongest; and 2) any premise which tries to resist, resolve, transform or heal it, etc., only serves to re-enforce and re-enact the False Core Driver because the False Core-False Self structure is holographic. All primary premises contain within them a self-fulfilling prophesy, namely, they prove themselves.

As stated previously, the False Core "I am worthless" has as its Nirvana, "If only I can become worthy and have value (through whatever means) I will reach merger with Mom (Nirvana) and be one again."

This way of proving worth is a goal which is never reached. For, as with all False Core-False Selfs, the obsessive-compulsive tendency to *right the wrong* (in the above case, to prove worth and value) through repetition compulsion (acting it out) only serves to re-play the same old story, thus, re-enforcing the False Core again and again.

THE REASONS FOR IDENTITIES

The story or reasons as to why I am (*fill in the blank*) comes in a later and different moment of space-time and hence is after the fact and only justifies and rationalizes what has already occured. In the above example, "I am worthless" because I have no value, I am too weak-willed, I am dependent, etc., are used by the False Core Driver to reinforce and justify its predicament.

Because this False Core believes it is worthless and it has no value, it might seek flattery or self-flatter to feel it is worth something. This self-flattery or creating an image to be flattered leaves this False Core-False Self with a sense of false pride, i.e., pride in their image. The False image creates False pride and is a made-up sense about oneself and one's accomplishments when there may not have been as any.[1]

THE I-DENTITY DISTRACTORS

To distract oneself from "I am worthless, I have no value," the focus is placed on feeling dependent, weak-willed or compensating by appearing over independent, self determined, valuable self-contained, totally together, and worthy. Unfortunately this is a *mask*.

---

[1] It should be noted that therapies which try to create a "good" self image over a "bad" one only re-enforce this cycle.

Chapter VII
FALSE CORE DISTRACTOR EMOTION
Self pride through False Images and through self flatery to overcome dependence.

THE OVERGIVER STRATEGY
This attempt to prove worth or value as an over-giver can be seen from several vantage points:

1. They can over-give to prove they have value or worth to overcome the False Core Driver.

2. They can *imagine* and create stories of overgiving, *believing* their made-up story when actually *they only give to get value*. In short, the False Core Driver of "I am worthless, I have no value" acts "as if" they were giving. But it is an act, done so that they can *get* something from you, i.e., a sense of value. They do get "value" from you, however, they know it was under false pretenses and that they are not worthy of it and, hence, they are left feeling, I have no value. In other words, *they give to get*. The double-bind is, if they do not "get," then they feel terrible, worthless with no value; and if they do get, they know it was under false pretenses so they cannot really take it in or it is short-lived and so they have to try to give more to get.

It should be noted that no matter how much they get, it is *not enough* to give them a sense of value. Why not? Because it is False giving, it is giving to get—not giving for giving's sake—and because it is driven by I have no value or worth. This leads to the distracting Identities of A) I am a giver or I will act "as if" I were a giver; B) I have to *get* or make up stories about something I have in order to have value; create an image of value and worth (wealth); C) create a False value, to look as if they have value, or D) Project their worthlessness on another so that they can "help" you and hence feel valuable.

3. In the final analysis, the *caretaker-overgiver is a defense* and is an age regressed child's attempt to merge with mom, defend

Since the False Core Conclusion is fused with the shock of the Realization of Separation it is part of the nervous systems' survival mechanism; the shock and the I am worthless, or any False Core must not under any circumstances happen again—hence the False Self solution.

Chapter VII

THE INSIDIOUS FALSE SELF

How sneaky and insidious can a False Self become in its defensive strategies?

Often times the only way this False Self, (*I have to prove I have value and worth*) can defend against the False Core of *worthlessness* is to project its own *unwillingness* to see and experience its own worthlessness onto another.

Because of the amnesic layer the False Self is unaware that it is projecting its False Core onto the "other." Soon the "other" is mistakenly perceived by this False Self as being unwilling to look at or acknowledge "their" stuff, (worthlessness).

This misperception affords this False Self mask several ways to defend against and resist the False Core of Worthlessness; 1) The False Self projects the disowned worthlessness (False Core) onto another; 2) It projects its *unwillingness* to look at their issues of worthlessness onto another, 3) It gives this False Self mask a *False Pride* for perceiving "others" issues of worthlessness which the "other" cannot see (because being unwilling to acknowledge issues belongs to this False Self projecting its False Core onto another—*It does not belong to the "other"*), 4) This False Self continually pathologizes and reminds the "other" (sometimes therapeutically, but always to defend and stay safe and in control of its issues of worthlessness which are part of the False Core-False Self complex). This serves two purposes. First it enables this False Self to feel *valuable* and second it enables this False Self to defend against *worthlessness*. 5) Hence, this False Self mask becomes driven to unknowingly justify its obsessive-compulsive drive to develop better and better skills to see (diagnose) "others" issues. This is due also to the False Core-False Self's deeply rooted unconscious transference-counter-transference age-regression. These skills which were acquired as a defense can yield *flattery* from *some* "others" for its False Self's skills. This False Self's obsessive compulsive defensive strategy is surrounded by amnesia. It originates as the age-regressed adult (via transference-counter-transference) attempts to control, figure-out, and manage another (mom/dad).

The tragedy of this integrated age regressed False Self mask is that it often creates and has a socially acceptable and appropriate context (sometimes professionally or socially) to unknowingly act-out and re-enforce the False Self's defensive strategies by compulsively unconsciously projecting and pathologizing "others", thus manufacturing an internal fantasy of "others" in need of their help because the "others" who received this projection cannot see it themselves (their stuff). Unfortunately, for this False Core-False Self complex some "others" find this integrated age-regressed projective tendency distasteful and want nothing to do with this False Self who continually insists on believing its projective identification is real. In this situation the "other" who finds this distasteful is labeled resistant by this False Self. This situation leaves the False Core-False Self Complex ultimately feeling even more worthless. Oddly enough, this False Self manages to find a context to "*get flattered*" and develop *False Pride* for their "diagnostic" or "healing" skills, all of which were cleverly designed by a False Self's desire to control, re-enforce, resist, survive and defend against the False Core of *worthlessness*.

against the narcissistic injury, the shock of the Realization of Separation and the False Core.

THE MAJOR DISTRACTING IDENTITIES:

I-dentity 1. I am dependent,
I-dentity 2. compensate and "*act*" over independent.

The over-giver can sometimes fuse with the dependent-over-independent (act) by attempting to make others dependent on them by making themselves indispensible to others. Actually they are extremely dependent but give an "air" mask of independence to cover the dependent age-regression.

Spiritual Practice Defense:
I must become worthy to (*fill in the blank*), merge,
feel **ESSENCE,** etc.

Oscar Ichazo uses the word *over*. Quantum Psychology uses the word over to describe a super compensation. It is a false pretend layer placed on top of the False Core, and separated by amensia. Furthermore, if in the case of "I am worthless, I have no value" they (the False Core-False Self) trys to get flattery and attention from the *external*, it is an attempt to heal the internal, thus it is a collapsing of the levels. They are trying to fix the internal "I am worthless" by going external (to get flattery). I once met an "I am worthless." who was in the Healing Arts. She exaggerated her worth by proclaiming that she could cure anything. If there was no one around to tell her how *worthy* and *valuable* she was, she would tell herself. So she developed *false pride* by believing her false overly-worthy/valuable story. In fact, she was so "into" flattery that if she felt depressed and you walked in and said, "Oh you look nice today," it would be like watering a flower that had been dry for six weeks.

Other strategies of this False Core-False Self is to project their False Self of overly-worthy onto another and then treat them "as if" this other were *so* special (worth), or project worthlessness (which, of course, the other cannot see) on another so as to not feel

Chapter VII

how worthless they feel and to feel really worthy and valuable for their "imagined" "fantasized" insights.

QUANTUM PSYCHOLOGY PRINCIPLE:
You must meet the problem at the level of the problem.

Many therapies and spiritual systems unknowingly resist, re-inforce or re-enact the False Core Driver. For example, if a client says "I'm worthless," the therapist wants to find out *why* they feel that way by taking a case history (story). This history is frequently a story used to justify their False Core position and give reasons, which were created after the fact to justify as to "why" they feel as they do. Therapists who believe and try to change the "personal history story" can pull a client away from the real issue by placing another layer on what drives their psychology (i.e., the False Core Driver). Then they get caught in the False Core's distraction which clients use to justify why they feel the way they do. This reinforces the False Core Driver. You feel the way you feel *because* of your *False Core*. Whatever happens *after* the False Core is solidified is used to *justify* the False Core. The *story* re-enforces or resists the False Core. Psychology is popular because it *re-enforces the story* and *resists the real issue*. For example, many therapies try to "help" you by getting you to com-pensate (False Self Compensator) for your False Core Driver or make it better, thus enhancing that which keeps the whole machine going.[2]

As another example, let us imagine a therapy which wants you to imagine the outcome of getting what you want at some future time. The only thing wrong with this method is: 1) the one doing the imagining is the False Self Compensator; 2) the False Core Driver-False Self Compensator is trying to overcome itself again and is re-enforcing itself; and 3) the therapy is re-enforcing the False Self Compensator's attempt to try to overcome the False Core.

Some therapies try to find solutions to problems. But the real problem is the False Core. When solutions are continually sought we begin to see a possible pattern (i.e., the problems→solutions *but* the

---

[2]It should be noted that therapies which try to "make it better" demonstrate an age-regressed therapist (Mom) saying (implying) to a client (child), "All better."

solution is placed on top of the problem→yet another needed solution. Recall a time you had a problem and created a solution, and the solution begot another problem which begot another solution. In this way the real problem of the False Core Driver is never addressed. The False Self Compensator-(AMNESIA)-False Core Driver is a holographic unit. You cannot have one without the other.

Other forms of therapy unknowingly try to change, reframe or reassociate the False Core Driver. Unfortunately, they are unknowingly enlisted in the aid of the False Self Compensator which again re-enacts the False Core-False Self.

Simply stated:
1. Trying to change the False Core Driver into something more positive only re-enforces it.
2. Reframing is a defense. It is done by the False Self and only adds another layer on top of the False Core Driver.
3. To make up or add new meanings, either consciously or unconsciously with hypnosis, only adds more to and hides the False Core Driver. This drives the False Core deeper and re-enforces the amnesic layer between the False Core Driver and the False Self Compensator.

QUANTUM PSYCHOLOGY PRINCIPLE:

Identities created after the False Core Driver try to get rid of, hide, heal, resolve or transform it, but can only, ultimately, reinforce it.

Quantum Psychology's False Core Driver:
I am Worthless, I have no value.

False Core Compensator:
I must prove worth so that I have value.

Quantum Psychology Distractor:
I-dentity #1—I am Dependent.
I-dentity #2 —I have to act over independent

I-dentity #1—You feel unconsciously worthless (deep down).
I-dentity #2 —I am so valuable, insightful, etc. I can help you to feel your value, and worth.

Quantum Psychology Distractor:
I-dentity #1—I am an overgiver (unconscious pretend)
I-dentity #2—I have to get (*fill in the blank*) to feel I am worthy or have value. "I must get flattery and attention, and tell myself how great I am, to have value."

Ichazo Identities:
I-dentity #1—"I am organized"
I-dentity #2—"I am disorganized"

PROTOCOLS FOR DISMANTLING THE FALSE CORE-FALSE SELF

The following protocol questions should be considered mini-meditations. The purpose is to ask a question and notice what, if anything, pops up, and then to acknowledge it and discard it. Please note this is not an analytic process wherein you try to figure out the answer to the question. It should be considered, as mentioned earlier, that the biological (nervous system's) scanning mechanism or psychological hypervigilance emerged very early to handle the chaos in one's childhood. For this reason typing, diagnosing, or analyzing is a subtle age-regressed defense against the chaos of the external context as it attempts to control through internal understanding the external. The questions can be seen as a flashlight. Their purpose is to get you to look at what's outside your awareness so that you can be free of it. To paraphrase Nisargadetta Maharaj:

> Anything you don't know about, you are a slave to. Anything you know about, you are free of. We enquire to become free of that which we do not know about.

The Way of the Human • The False Core and the False Self

FALSE CORE DRIVER—I AM WORTHLESS

1. Who modeled the *I am worthless or I have no value" I-dentity*? Scan your body and notice where this identity is.

2. Where in your body do you experience this *I am worthless and have no value I-dentity*? Ask yourself this question and write down the answer until nothing pops up.

Therapeutic Note

At this point it is advantageous to externalize the *I*-dentity. This can be done in the following way:

Step I: Notice where in the present time body the *I*-dentity is located.

Step II: Notice the *I*-dentity's size and shape.

Step III: Take the label off of the *I*-dentity. Have it as energy.

Step IV: Allow the *I*-dentity as energy to move from the present time body to another physical location in the room, and then to become solid again.

Step V: Continue the rest of the enquiry protocols, asking the *I*-dentity the questions, and allowing the *I-dentity to respond.*

3. Has this "worthless" or *"I have no value" I-dentity* ever pretended to be worthless/no value in order to be taken care of? Ask this *I*-dentity this question and write down the answer until nothing pops up.

4. Has this *"I am worthless"* or *"I have no value" I-dentity* ever tried to control another person by making them dependent on

Chapter VII

it? Ask the *I*-dentity this question and write down the answer until nothing pops up.

5. If you take on the *"I am worthless"* and/or *"I have no value" I-dentity,* do you feel age-regressed, and if so, to approximately what age? Ask the *I*-dentity this question and write down the answer until nothing pops up.

6. Does this *"I am worthless," "I have no value" I-dentity,* ever create fantasies of times in the future when someone will come and take care of it? Ask the *I*-dentity this question and write down the answer until nothing pops up.

7. Has the *"I am worthless, I have no value" I-dentity* ever been used to justify in some spiritual way, like being worthless with a teacher, a guru, or God or making worthlessness a noble trait. If yes, what spiritual philosophies and disciplines has the False Core taken on to justify its worthlessness or dependency? Ask the *I*-dentity this question and write down the answer until nothing pops up.

8. Has the *"I am worthless, I have no value" I-dentity* ever justified its avoiding of taking care of its health? Ask the *I*-dentity this question and write down the answer until nothing pops up.

9. Has the False Core ever sought attention to avoid feeling the *"I am worthless, I have no value" I-dentity*? Ask the *I*-dentity this question and write down the answer until nothing pops up.

10. Has this False Core ever imagined that if it were worthless that someone would feel sorry for it and try to save it and merge with it? Ask the I-dentity this question and write down the answer until nothing pops-up.

11. Has this False Core ever used the *"I am worthless, I have no value" I-dentity* as a way or justification to not manage its own

finances? Ask the *I*-dentity this question and write down the answer until nothing pops up.

12. During the taking on of this *"I am worthless, I have no value" I-dentity*, was there any chaos that occurred? Ask the *I*-dentity this question and write down the answer until nothing pops up.

13. During the taking-on of this *"I am worthless, I have no value" I-dentity*, what did the *I*-dentity have to pretend that it didn't understand? Ask the *I*-dentity this question and write down the answer until nothing pops up.

14. During the taking-on of this *"I am worthless, I have no value" I-dentity*, what beliefs, assumptions or decisions did this *I*-dentity make? Ask the *I*-dentity this question and write down the answer until nothing pops up.

15. During the taking-on of this *"I am worthless, I have no value" I-dentity*, what future catastrophic or pleasant fantasies did it create? Ask the *I*-dentity this question and write down the answer until nothing pops up.

16. Notice where in your physical body you experience the *"I am worthless," "I have no value" I-dentity*.

17. Peel back the layer called *"I am worthless, I have no value" I-dentity*, and go into the **SPACIOUSNESS** inside the physical body.

18. From inside the **SPACIOUSNESS** inside the physical body, how does the *"I am worthless, I have no value" I-dentity* seem to you now?

19. From inside the **SPACIOUSNESS** inside your physical body, ask the (mis)labeled emptiness and the *"I am worthless," "I have no value I-dentity,"* "What are you seeking more than any-

Chapter VII

thing else in the world?" Allow them to answer and write down the answers.

20. From inside the **SPACIOUSNESS**, experience the essential quality which the (mis)labeled emptiness and the *"I am worthless"* and/or *"I have no value"* I-dentity was seeking.

21. Notice the size and shape of (mis)labeled emptiness and the *"I am worthless, I have no value"* I-dentity.

22. See the spaciousiness, the (mis)labeled emptiness and the *"I am worthless, I have no value"* I-dentity—as made of the same substance as the **SPACIOUSNESS**—while you continue to experience the essential quality that the emptiness label and the *"I am worthless, I have no value"* I-dentity was seeking

FALSE SELF COMPENSATOR- THE "I MUST PROVE MY WORTH— I MUST PROVE I HAVE VALUE I-DENTITY

1. This *"I must prove my worth"-"I must prove I have value"* I-dentity, where do you experience it in your body? Write down the answer until nothing pops up.

2. Notice who modeled this *"I must prove my worth"-"I must prove I have value"* I-dentity.

Therapeutic Note
At this point it is advantageous to externalize the *I*-dentity. This can be done in the following way:

      Step I: Notice where in the present time body the *I*-dentity is located.

      Step II: Notice the *I*-dentity's size and shape.

Step III: Take the label off of the *I*-dentity. Have it as energy.

Step IV: Allow the *I*-dentity as energy to move from the present time body to another physical location in the room, and then to become solid again.

Step V: Continue the rest of the enquiry protocols, asking the *I*-dentity the questions, and allowing the *I-dentity to respond.*

3. Notice the approximate age that you age-regress to in order to take on this *"I must prove my worth"-"I must prove I have value" I-dentity*. Write down the answer.

4. Notice how old you feel when you take on this *"I must prove my worth"-"I must prove I have value" I-dentity*. Write down the answer.

5. Notice that the *"I must prove my worth"-"I must prove I have value" I-dentity* is actually a cover, a False Self or mask used to hide the *worthless—no-value I*-dentity.

6. Has this *"I must prove my worth"-"I must prove I have value" I-dentity* ever been used as a way of resisting acknowledging dependence? Ask the *I*-dentity this question and write down the answer until nothing pops up.

7. Has the I-dentity ever felt that if it "acted" or proved worth, people would want to merge with it? Ask the I-dentity, this question and write down the answer until nothing pops up.

8. Has this *"I must prove my worth"-"I must prove I have value" I-dentity* ever been used as a way of keeping its own separateness from another and having a separate self? Ask the *I*-dentity this question and write down the answer until nothing pops up.

Chapter VII

9. Has this *"I must prove worth"-"I must prove I have value" I-dentity* ever created an image or appearance for others to admire? In this way, get attention or flattery so that it feels like it has some value or worth. Ask the *I*-dentity this question and write down the answer until nothing pops up.

10. Has the *"I must prove my worth"-I must prove I have value" I-dentity* ever been used to *act* independent to prove or act "as if" it didn't need anybody or anything as a way to hide worthlessness or dependency. Ask the *I*-dentity this question and write down the answer until nothing pops up.

11. Has the *"I must prove my worth"-I must prove I have value" I-dentity* ever been used as a defense against feeling needy, needing another or feeling like a bottomless pit of needs? Ask the *I*-dentity this question and write down the answer until nothing pops up.

12. Has the *"I must prove worth-I must prove I have value" I-dentity* ever been used to avoid feeling physical illness? Ask the *I*-dentity this question and write down the answer until nothing pops up.

13. Has the *"I must prove my worth—I must prove I have value"* I-dentity ever been used to avoid looking, showing, acknowledging illness or that *anything* could be wrong. Because illness or *anything* not being great means "I am worthless." Ask the *I*-dentity this question and write down the answer until nothing pops up.

14. Has the *"I must prove worth-I must prove I have value" I-dentity* ever been used as a motivator to become independently wealthy or to act "as if" I don't need money because "I have value." Ask the *I*-dentity this question and write down the answer until nothing pops up.

The Way of the Human • The False Core and the False Self

15. Has the *"I must prove worth-I must prove I have value"* *I*-dentity ever been used as a way to flatter yourself or get flattery for your skills so that you would feel like you had value. Ask the *I*-dentity this question and write down the answer until nothing pops up.

16. During the taking on of this *"I must prove worth- I must prove I have value" I-dentity*, was there any chaos that occurred? Write down the answer until nothing pops up.

17. Notice how the *"I must prove my worth"-"I must prove I have value" I-dentity* is truly a False Self and how it masks the *"I am worthless" I-dentity* and how it is a *false* sense of worth.

18. During the taking on of this *"I must prove worth-I must prove I have value" I-dentity*, what decisions, assumptions or beliefs did this *I*-dentity create? Ask the *I*-dentity this question and write down the answer until nothing pops up.

19. During the emergence of this *"I must prove worth-I must prove I have value" I-dentity*, what fantasies did this *I*-dentity create regarding its grandiosity? Ask the *I*-dentity this question and write down the answer until nothing pops up.

20. This *"I must prove my worth-I must prove I have value" I-dentity*, what does it say to itself to make itself appear to have more worth and value? Ask the *I*-dentity this question and write down the answer until nothing pops up.

21. This *"I must prove my worth-I must prove I have value" I-dentity*, what stories of greatness does it make up to reinforce its existence as a separate self? Ask the *I*-dentity this question and write down the answer until nothing pops up.

22. Where in your body do you experience the **SPACIOUSNESS**? Write down the answer until nothing pops up.

Chapter VII

23. Notice how the *"I must prove my worth-I must prove I have value" I-dentity* labels that **SPACIOUSNESS**. For example, does it label it as empty, "dependent," "worthless, etc.?" Write down the answers until nothing pops up.

24. Peel back the layer or (mis)label of the emptiness and *"I must prove my worth-I must prove I have value" I-dentity* and go inside the **SPACIOUSNESS** in the physical body.

25. How does the *"I must prove my worth"—"I must prove I have value" I-dentity* and the (mis)labeled emptiness seem to you from back there in the **SPACIOUSNESS**? Write down the answers.

26. Ask the *"I must prove my worth-I must prove I have value" I-dentity* and the (mis)label of emptiness from "inside" the **SPACIOUSNESS**, "What are you seeking, more than anything else in the world?" Allow *them* to answer and write down the answers.

27. Experience that essential quality that the identities and the label are seeking.

28. Notice the size and shape of the emptiness (mis)label and the *"I must prove my worth"-"I must prove I have value" I-dentity*, and the **SPACIOUSNESS** surrounding it.

29. See the **SPACIOUSNESS**, the label and the I-dentity as the same substance, as you continue to experience the essential quality of **ESSENCE**.

DEMONSTRATION—
FALSE CORE DRIVER DISTRACTORS→
FALSE CORE DRIVER

The demonstration takes place with Barbara, a 49-year old therapist from Texas. In this session, we focus first on both the distractors and how they relate to and lead to the False Core Driver.

Wolinsky: The over independent identity, where do you feel it in your body?

Therapeutic Note
The False Core driver as well as the distractor identities are body centered because it is a conclusion drawn from the shock of the separation.

Barbara: My shoulders, my head. The image I have is like a water buffalo with one of those yokes on it.

Wolinsky: Who modeled this over-independent identity for you?

Therapeutic Note
The False Core (often) has a model. It is often "taken on" as the only way to have a relationship (with mom) and avoid the shock of the Realization of Separation. This is why people who look for models, formulas, for life are (often) age-regressed. They are re-enacting by trying to overcome the shock of the Realization of Separation by looking for an exterior model or "how to, be, do, act, feel, etc., "as if" there was a book of rules; what to do when (*fill in the blank*) occurs or if (*fill in the blank*) occurs.

Barbara: My mother. Although I did it better than my mother did it.

Therapeutic Note
Here she "takes on" the distraction that mom uses and then resists it.

Wolinsky: What did you assume, decide or believe that got you to do it better than her?

Barbara: Well that she was weak willed, and I was stronger willed.

Chapter VII

Wolinsky: What age approximately were you when you took on this?

Therapeutic Note

It is important sometimes to see if there is a traumatic memory which would make the energy bound up in the False Core stronger.

Barbara: I am the youngest co-dependent on record. I was 11 months old.

Wolinsky: 11 months old when you became co-dependent?

Barbara: Yes. There's stories of me crawling into my sister's room, pulling myself up on her crib and helping her go to sleep so my mother could sleep. She was sick. That's when I had a sense of it in my body that young. Again, this pulling up has to do with all in here (points to chest).

Wolinsky: Pulling up on the bars?

Barbara: Yeah, pulling up on the crib.

Wolinsky: If you go into that image of you pulling up on the crib, what thoughts, feelings, fantasies come up for you?

Barbara: I'm going to cry. I feel like I can't breathe.

Wolinsky: Feel that for a moment.

Barbara: I am also aware of the other half of my body not being able to walk because I couldn't walk yet. I feel real split.

Wolinsky: Between?

Barbara: I can't walk and super strong willed. Willful. Head strong.

Wolinsky: How does your jaw feel?

Therapeutic Note
"I" could see the strain in her jaw.

Barbara: I'm clenching my teeth. Stubborn, persistent, in opposition to those whose wishes or commands ought to be respected or obeyed.

Wolinsky: So where in your body do you experience the dependent or even worthless identity?

Therapeutic Note
Here we move to the other distractor to the False Core of "worthless," "I am dependent."

Barbara: I would say from my stomach down. This is a layer. This independent is a layer. The dependent is underneath it. It is hard. I feel it in my jaw and in my shoulder blades.

Wolinsky: Has the overly independent, willful I-dentity ever been used as a way of resisting or not acknowledging dependence?

Barbara: Yeah.

Wolinsky: Give me an example of one.

Therapeutic Note
Always ask for a specific example.

Chapter VII

Barbara: I took care of a lot of kids in my family so I became this big sister/mother figure. I kept that illusion going strong, the big sister, and didn't get to feel what I needed.

Wolinsky: So you didn't get to feel dependent?

Barbara: No.

Therapeutic Note
There is a resistance to feeling dependent (which an infant is) so we want to draw out the resisted experience.

Wolinsky: Can you feel dependent right now?

Barbara: Yeah, I can feel it. I feel very tremulous.

Wolinsky: Stay with tremulous as you go into that kind of dependency. Feel the tremor.

Barbara: It gets into my hands and arms. I can feel it in my legs. Like wobbly. Like that baby that didn't know how to walk. It has that real unsureness to it.

Therapeutic Note
It is important that she begins to be willing to experience the dependent which is unacknowledged.

Wolinsky: I would like for you to make a statement, "I don't know how to walk but I have to take care of the new baby."

Barbara: "I don't know how to walk, but I have to take care of my mother." I feel like I didn't get to learn a lot of things because I had to take care of my mother.

169

Wolinsky: Go into the dependent/shaky, just for a moment, and then to not feel that intentionally, go into the distractor I-dentity of *acting* independent and tighten down the shaky dependency.

Barbara: Okay.

Wolinsky: Do that again.

Barbara: Okay.

Wolinsky: Now, create it.

Barbara: Okay.

Wolinsky: Now, create it again.

Barbara: Okay.

Wolinsky: Now, stop creating it.

Barbara: Okay.

Wolinsky: Now, create it.

Barbara: Okay.

Wolinsky: Now, stop creating it.

Barbara: Okay.

Wolinsky: How are you doing?

Barbara: Much clearer.

Chapter VII

Therapeutic Note:
I am asking her to knowingly, consciously intentionally to go into and then create the pattern she is unknowingly, unconsciously unintentionally creating. And I will begin to have her notice that the over (false) independent I-dentity is really a *false* over-independence.

Wolinsky:	Have you ever used this overly (false) independent, willful I-dentity as a way of keeping yourself separate from another, (like your mother)?
Barbara:	Well, Yeah, because it's like a hiding place from the pain of the separateness.
Wolinsky:	Now, take off this overly independent identity like clothing and put it over there (the other side of the room) and be in the dependent identity, the one with needs, and now look at me, and let me know when you are "there."

Therapeutic Note
Always externalize the Identities from the present-time body. This helps them to "get" the Identities are not them.

Barbara:	They are over there.
Wolinsky:	Now, if you caught an image of your mother out of the peripheral vision, could you stay in relationship with her and be dependent?
Barbara:	No.
Wolinsky:	So the only way you can have a relationship with your mother is by acting *over independent* and having basically no needs. Is that correct?
Barbara:	Right.

Wolinsky: Okay. Have an image of your mother over there (other side of the room) and I want you to say to your mother, "Mom, the only way I can have a relationship with you is by pretending *(fill in the blank)*.

Therapeutic Note

I am using incomplete sentences to make the implicit explicit, i.e., acknowledge the unacknowledgeable.

Barbara: (To Mom)

"The only way I can have a relationship with you is to *pretend I am strong willed.*"
"The only way I can have a relationship with you is to *pretend that I don't need you.*"
"The only way I can have a relationship with you is to *pretend that I have it all together.*"
"The only way I can have a relationship with you is to *pretend that I know everything.*"
"The only way I can have a relationship with you is to *pretend I haven't been figured out.*"
"The only way I can have a relationship with you is to *pretend I don't have feelings.*"

Wolinsky: Have you ever used this overly independent (distractor) identity to create an image or appearance for others to admire?

Barbara: Well, with my siblings.

Wolinsky: How about your mom?

Barbara: And my mom. I went to college. I'm a successful person.

Wolinsky:	Have you ever used this overly independent willful (distractor) identity to prove that you didn't need anybody or anything?
Barbara:	Well, I think in relationships with men, I have done that. Get in relationships where I don't get any of my needs met and act like I can survive that.
Wolinsky:	Did you ever project your needs or dependency on a man, then take care of them?

Therapeutic Note

Here we are having her talk about the *false* independent (no needs distractor I-dentity) and chosing a partner to "act out the other side of the I-dentity."

Barbara:	Oh, yeah, always.
Wolinsky:	What do you feel right now?
Barbara:	To not have to do that? Oh, man, it is so freeing. It's kind of scary too because it's like who am I? I am in that zone. If I'm not overly independent or dependent, where and who am I?

Therapeutic Note

There is a lot of energy and attention placed on the distracting Identities. When they begin to dissolve, there is oftentimes an inability to locate yourself in space/time because the distracting I-dentity had been the way you organize the "you" you call yourself in space/time and is your space/time reference point to distract against the False Core driver.

Wolinsky:	Have you ever used the overly independent willful identity as a defense against feeling needy or

needing another as a way of not having to feel the worthlessness?

Therapeutic Note

Here we are peeling back the distractors to get to the False Core driver of "I am worthless."

Barbara: Well, I think in my family. The big sister, strong person. Then I don't have to need anybody. I just give. I am incapable of doing it really, so I feel like I really do not have any value or worth *really*.

Wolinsky: Tell me a difference between you and the identity.

Therapeutic Note

Since she is not her False Core, we want to separate her from it. Nisargadetta Maharaj had several favorite questions/statements:
1. Tell me a difference between "you" and "it."
2. Who came first, "you" or "it?"
3. Prior to (*fill in the blank*), stay there.

Barbara: Well, I don't feel this thing. I feel like I am in my body more. Like I am here more. And the strength is a different feeling. It's not this yoke thing. It's more subtle, more alive. I'm different.

Wolinsky: Now, feel that difference and have an image of your mother over there and look at your mother *now*.

Barbara: Uh huh, I can.

Wolinsky: How does she seem to you *now* and how do you feel right *now* looking at her?

Chapter VII

Therapeutic Note

I am emphasizing *now* to differentiate from *then* (the age-regressed little girl).

Barbara: I feel sad.

Therapeutic Note

Notice how she is age-regressed. I differentiate her present time "self" from her age-regressed little girl I-dentity.

Wolinsky: Where in your body do you feel the little girl who was worthless and dependent?

Barbara: In my heart.

Wolinsky: Tell me a difference between you and the little girl I-dentity.

Barbara: I am here now and she is there.

Wolinsky: Put the litlle girl "over there" (other side of the room) with mom. Take the little girl off of your body and notice the relationship between the two Identities and "take on" the little girl I-dentity. Looking at your mother right now, could you say to her, and complete this sentence, "Mom, I couldn't show you my (*fill in the blank*).

Barbara: "Mom, I couldn't show you *who I really was*."
"Mom, I couldn't show you *my needs*."
"Mom, I couldn't show you *how much I needed you*."
"Mom, I couldn't show you *how pissed I was that you made the choices that you made*."
"Mom, I couldn't show you *the truth of what was going on*. That's real frustrating."

Wolinsky: "Mom, what I wanted and didn't get was (*fill in the blanks*)."

Barbara: "Mom what I wanted—and didn't get—was for you to tell the fuckin' truth about what was going on in our family and then do something about it and to care about me in that truth. I see how it was impacting me. Yeah, that's what I wanted. Mom, what I wanted was for you to see how it affected me, not just you all the time. Yeah, that feels good.

Therapeutic Note

The most common question asked me in seminars is, "Is it a necessary step to create new beliefs or decisions? Quantum Psychology does not *create, reprogram, re-decide or re-associate* anything new. In Quantum Psychology, this new decision or belief is seen as *adding another layer and as a defense* against the False Core Driver and the shock of the Realization of Separation.

However, if a person is acting out of an age-regressed I-dentity, the age-regressed identitiy has unacknowledged wants, needs, emotions, information, identities, etc. They must first be acknowledged make the implicit-explicit. Simply put, I first have to have a *whole* person. A whole person is someone who has acknowledged and made the implicit-explicit, but realizes the "unwhole person who has unacknowledged wants, etc. is age-regressed. Why? Because if you were totally NOW there is no issue. Then, the person is naturally whole and you can go beyond. Trying to go beyond your personality before you have acknowledged it, only aids in re-enforcing the personality and its defenses. For example, if a person who has not acknowledged their age-regressed psychology tries to go beyond into **ESSENCE** and **VOID**, they only create more pain. Why? Because the new philosophy and techniques are placed on top of the False Core Driver and act as more fuel to keep the False Core driver-False Self Compensator machine going. The False Core driver must be acknowledged, accepted and seen for what it is: *the personal "I" you call yourself*. Then you can go beyond. For example, there is an Indian teacher and student of Nisargadetta Maharaj who tells people, "You are not the doer." Now, this teaching has many levels to it. If you tell a person whose False Core-False Self is still operating this, then they will use it to re-enforce their False Core-False Self.

In Quantum Psychology not only do we want to meet the problem at the level of the problem, but we want to meet the student at the level of the problem the student is having. Therefore, if a student is in the False Core and "thinks" they are unwhole, first help them to get the unacknowledged experiences, dismantle the identities (age-regression), then they can move beyond. To try to give a new philosophy to a False Core-False Self, age regressed person, is dangerous to them because the new philosophy will be placed on top of the False Core and it will reinforce their False Core-False Self, and hence, their pain.

Wolinsky: So you had to suppress your needs in order to organize around her?

Barbara: Needs and truths. They kind of go together.

Wolinsky: How are you feeling now?

Barbara: I feel like I can see. Like a cataract just got cleared or fog cleared. I can see clearer.

Wolinsky: If you fused together the world and mom, if they were fused together, what are you creating?

Therapeutic Note
Most people unknowingly fuse together mom or dad with the world or spirituality. The six step de-fusion process is used to unfuse the trance-ference equals the world, etc., and dismantle it. (See Six Step Fusion-De-Fusion process.)

Barbara: Oh, yes, a workaholic

Wolinsky: What did you assume, decide or believe that got you to creating that?

Barbara: More this, more false independence.

Wolinsky: And what did you assume, decide or believe that got you to create more false independence?

Barbara: That was just going to be the way to feel okay. Like it was the way.

Wolinsky: The way of the truth and the light. How does that seem to you now?

Barbara: Well, the fusion with mom and the world just seems like a little, teeny room, like, Oh, my God.

Wolinsky: Are you still creating that right now?

Barbara: Yeah.

Wolinsky: If you fused together mom and the world and they are fused together, what are you not creating?

Barbara: Space, expressions, self-expression.

Wolinsky: What did you assume, decide or believe that got you not to create space and self expression?

Barbara: Well, that it would lead to needs. It would lead back to feelings and needs. You can't go there. I can't go there.

Wolinsky: If you fused together mom and the world, what are you resisting?

Barbara: Myself.

Wolinsky: What did you assume, decide or believe that got you to resisting yourself?

Chapter VII

Barbara: Your taking me down that path Stephen.

Wolinsky: The one you don't want to go down. (Laughs)

Therapeutic Note
She is correct. "I" am guiding her on the path to her Organizing Principle the False Core Driver. "I am worthless" like a needle on the groove on a record.

Barbara: You see all that stuff keeps me away from that. I can feel it. My existence is not valued.

Therapeutic Note
No value is often the same as "I am worthless."

Wolinsky: Yes.

Barbara: So I can just feel that whole false independent act is to resist going into no value.

Wolinsky: Into the false construct called. . . ?

Barbara: Well, all the identities. Well, there is the false construct that *"I'm worthless."* For me, it's more like *not valued. No value.* That's the word. *No value.* Because if you were valued, you would be interacted with.

Wolinsky: Since your mother didn't interact with you, therefore, you had *no value.* Notice the size and shape of that belief.

Barbara: Huge and big. Like the campus.

Wolinsky: Step into the belief and see the world through it.

Barbara: It's lax, scarcity, not enough.

Wolinsky: Merge with it and view yourself through that lens?

Barbara: A speck. A germ. Can't even see me. That's worthless. Germs are pretty powerful, it's even worse than that. *No value.*

Therapeutic Note

The False Core is a lens through which you view the world, through which you view yourself and through which you imagine the world views you.

Wolinsky: If you fused together mom and the world, what are you resisting?

Barbara: That experience (no value).

Wolinsky: What did you assume, decide or believe that got you to resist that experience (no value)?

Barbara: That experience (no value) was true.

Wolinsky: Oh, okay, so since no value was true, and such an awful experience, you need to resist it.

Barbara: Right. And then if I resist it, it won't be true.

Wolinsky: Where in your body do you feel the belief, "If I resist it, it's not true?"

Barbara: In my chest.

Wolinsky: Take the label off. Have it as energy. Now, that experience called no value, create that experience, that energetic experience, create it the size of the room.

Chapter VII

QUANTUM PSYCHOLOGY PRINCIPLE:
The greater the ability to create and then vary the experience, the more the resistance to "it" disappears and you are willing to *have* it. Hence, the greater the subjective experience of freedom.

Barbara: Like big smallness. It's emptiness. It is like nothing.

Wolinsky: Now, create it the size of Washington. Now, create it the size from Seattle to San Diego. Now, create it from the size of California, the western seaboard, all the way to Mississippi.

Barbara: I mean, I could fill it everywhere.

Wolinsky: Then, make it as big, include South America, make it as big as South America. Now, go Pacific, Japan, Hawaii.

Barbara: I can feel it everywhere. I feel it all over the planet. I can look down and see the earth.

Wolinsky: Good. Now, create it the size of the solar system. Now, create it the size of the galaxy.

Barbara: It's getting absurd.

Wolinsky: Now, create it the size of the universe. Now, take the label off of it and have it as energy.

Barbara: It's more peaceful the farther out the galaxy it gets.

Wolinsky: Yes. How are you doing now?

Barbara: I feel here. There is a light shifting in the eyes. Like in the eye doctor.

Therapeutic Note
Anything that happens to you, if you knowingly, consciously, intentionally create it, the automatic creation of it dissolves.

Wolinsky: Now, if you separate mom and the world, and if they are separate, what gets created?

Barbara: Well, room for me. I feel like I can be here.

Wolinsky: Now, if you separate mom and the world, what are you *not* creating?

Barbara: Tightness, constriction.

Wolinsky: And if you separate mom and the world, what, *if anything*, are you resisting?

Barbara: Nothing, very clear and quiet.

Wolinsky: If you fuse together needs and feelings equal no value and unworthy—

Therapeutic Note
This is an important fusion because every time a need or feeling comes up, the False Core is ignited, i.e., feelings = False Core driver.

Barbara: Unworthiness.

Wolinsky: What did you assume, decide or believe that got you to creating worthlessness?

Barbara: Needs equal no value.

Wolinsky: Can you say a little more?

Barbara: I get to hang out in worthlessness instead of knowing my needs. To me, needs and being with your-

Chapter VII

self are what go with the package. You are in the body. Needs means "I am worthless."

Wolinsky: If you fuse together needs equals no value, what are you *not* creating?

Barbara: Not creating, self exploration or creativity.

Wolinsky: Okay, Yes. What did you assume, decide or believe that got you to not creating self exploration or creativity?

Barbara: Again it would take you someplace dangerous, like into *worthlessness*.

Wolinsky: Needs and creativity bring you to worthless. If you fuse together the idea that if needs equals I have no value and I am worthless, if that's all fused together, what are you resisting?

Barbara: My anger.

Wolinsky: What did you assume, decide or believe that got you to resisting your anger about it?

Barbara: Well, anger is bad, anger is scary, anger is another need, another expression. It comes back to need.

Wolinsky: So obviously if you had that then it would automatically mean you are worthless.

Barbara: Right. So every time I have a feeling, I am worthless. If I have a need, I'm worthless.

Wolinsky: Yes. Say that again.

Therapeutic Note

Making the Implicit-Explicit.

Barbara: Every time I have a feeling, I am worthless. Every time I have a need, I'm worthless.

Wolinsky: I understand. When you had a need, or when you had feelings, did your mother give you this worthless energy?

Barbara: Uh huh, I was shamed badly. Disgust.

Therapeutic Note

Force Theory says a force comes at you and you counter it in some way.

Wolinsky: So this energy coming at you, this disgust (force).

Barbara: That's in there. That's in between the independent/dependent.

Wolinsky: If your mother is over there, I want you to imagine this energy called disgust coming at you and then you creating in response that "I am worthless," shutting off your needs and feelings.

Therapeutic Note

Children pick up, "take on" and fuse with mom's energy.

Barbara: Oh, Yeah, it's there.

Wolinsky: Now, what I would like for you to do as the energy of mom comes at you called disgust is to allow that energy just to go back to mom.

Therapeutic Note

Take the label off and allow the energy to go back to its source.

Wolinsky:	Notice where in your body you feel this "self hate."
Barbara:	In my throat.
Wolinsky:	Is it your self hate, or is it the hate of mom to or for you for having needs, creativity or being.
Barbara:	It is my mother's hate.
Wolinsky:	Okay, take the label off and allow it to go through the little girl identity and then from her back to mom.

Therapeutic Note

No **ESSENCE-I AM** hates or "beats up themselves" or is critical of itself for (*fill in the blank*). It is a fused internalized parent. This is a *highlight* in Quantum Psychology. *This is not a critical part of you. It is a fused parent who is not you. Do not re-frame it, give it a new label as a part of you.* Nisargadetta Maharaj said, "What you know about yourself came from outside of you. Therefore discard it." Quantum Psychology is doing just that through the context of Western psychology. The energy goes back to mom and hence it is discarded.

Barbara:	I can see a wave, like I was at the beach this morning, you just take that step and it just goes back, you miss it.
Wolinsky:	How do you feel now?
Barbara:	Well, I feel separate from it. That feels good. I feel good. I feel really present.
Wolinsky:	If you separate having needs equals no value, what are you creating?
Barbara:	The word creativity comes to mind.

Wolinsky: If you separate needs and no value, what are you not creating?

Barbara: Self hate.

Wolinsky: If you separate needs and no value, what, if anything, are you resisting?

Barbara: Nothing.

Wolinsky: How do you feel?

Barbara: I feel centered, clear, quiet.

Wolinsky: So would it be fair to say that you saw yourself through your mother's eyes.

Barbara: And through her mother and her mother's mother.

Wolinsky: Can you say that to me?

Therapeutic Note

She sees herself through the False Core of mom and the (image) of mom. This will be discussed in greater detail in the next chapter.

Barbara: I see myself through my mother's eyes, and my grandmother's eyes.

Wolinsky: From where do you see yourself?

Barbara: From where? Well, of course, outside of myself.

Wolinsky: From where do you see yourself? Tell me a location.

Barbara: About here. (A few feet away) Looking back at me.

Chapter VII

Therapeutic Note

She sees herself, from outside of herself, just as her mother who was outside of her, saw her.

Wolinsky: Good. Tell me another location from where you see yourself.

Barbara: Yeah, it's an external reference kind of thing.

Wolinsky: Now, just for fun, see yourself through your mother's eyes.

Barbara: Okay.

Wolinsky: Now, let go of that and see yourself as yourself.

Barbara: Like from the inside.

Wolinsky: Yeah.

Barbara: It's a lot more comfortable. I feel lighter.

Wolinsky: What I would like for you to do is have your mother's eyes again looking at you.

Barbara: Okay. It really is over there.

Wolinsky: Now, take those eyes, and move them over there (another location in the room) and look at you through your mother's eyes, but from over here. (another place in the room)

Therapeutic Note

We want to get her to loosen her frozen mother's eyes, which she sees herself through and which were trance-ferred onto others.

Barbara: Yeah, that starts to free it up a little because It's really in place over here.

Wolinsky: Now, take and see yourself through your mother's eyes, way over there. (Yet another place.)

Barbara: It's a comfort. It's breaking it up. It seems more false over there.

Wolinsky: Now, take your mother's eyes that are there and rather than create them, actually take them and put them over there (another place in the room).

Barbara: I'm less susceptible over there.

Wolinsky: Now, take them and put them over there (another place).

Barbara: When I look back at myself, I can't do it as well.

Wolinsky: Now, take those eyes, and put them over by the sign No Food or Drink allowed in this Room.

Barbara: Yeah, no needs. (Laughs)

Wolinsky: Good. Now, take it and put it behind your head.

Barbara: It feels good. It feels like aerobics or something. Like I'm stretching out or something.

Wolinsky: Stretch them out like a rope, like here are the eyes, there are all of these eyes. Now, skip rope with it.

Therapeutic Note

Again, more variability, greater freedom over the automaticity.

Barbara: I can get into it.

Wolinsky: How do you feel?

Barbara: That was fun. It's like I love the irreverence of it because this is a very reverent reference in here. It is like a scared place inside me.

Wolinsky: How do you feel knowing that you took on your mother's unworthiness?

Barbara: Well, I just feel like it is more hers because I don't feel all hung up about it or angry.

Wolinsky: How do you feel in your core (biological)?

Barbara: Clean, clear. I'm resisting that there is some pain about aliveness equals worthlessness. I'm resisting feeling the pain around aliveness equals worthlessness.

Wolinsky: When the aliveness comes up and you have to squish it down, it creates the experience of pain. So feel the aliveness and intentionally squish it down.

Therapeutic Note

I am asking her to do (squish the aliveness) knowingly, consciously, intentionally what she is doing *un*knowingly, *un*consciously, *un*intentionally.

Barbara: Right, okay right. I've got it. The pain is the shutting off of aliveness.

Wolinsky: So, intentionally experience the aliveness and intentionally push it down.

Barbara: It's like resignation.

Wolinsky: So, intentionally experience the aliveness and push it down. Take the label off of both of them and have them both as energy and allow them to do whatever they do.

Therapeutic Note

Always do a six-step de-fusion/fusion process. In this case, aliveness = worthlessness.

Wolinsky: So you had a force coming at you called the squisher (mom). And you created *no value, worthless* in response.

Therapeutic Note

Force theory suggests that a force comes at you *and* to counter it, you create something.

Barbara: My internal experience. My parents did something but not about me inside.

Wolinsky: Your internal experience was no value. Tell me something you decided about that?

Barbara: I decided that if I had any needs around, I was needy. Like you just don't have feelings about this. It was right and wrong. You go home and say it was wrong and then they would take this guy to court. But there's no feelings about it. So the feelings about it, I decided I was sick. Something was wrong with me because I had feelings about it. Does that make sense?

Wolinsky: Yes. Tell me something else you decided?

Barbara: I decided I was crazy.

Wolinsky: Are you still deciding that?

Chapter VII

Therapeutic Note

Decide is a moment frozen in space-time. When you use *ing* it makes a decision a process. It gives the decision motion.

Barbara: Uh huh.

Wolinsky: Right now?

Barbara: No. A little bit. I still have that.

Wolinsky: Good intentionally decide it.

Barbara: And crazy means to have feelings and needs. And then you are worthless and then you are in the nursing home.

Wolinsky: Notice the size and shape of the belief called "I am crazy, I must be crazy."

Barbara: That's big. Maybe as big as these two rooms.

Wolinsky: Notice the **EMPTINESS** that it is floating in.

Barbara: Uh huh.

Wolinsky: See the **EMPTINESS** and belief as the same substance. How are you doing now?

Barbara: I don't feel like I want to bolt. I feel like I am more here.

Wolinsky: So I want you to notice the associations. And how, every time a feeling comes up, a chain of associations leads into *worthlessness*. If a need comes up it leads to, or is equal to, worthlessness. If aliveness comes up, it equals worthlessness. How are you feeling now?

Barbara: I don't feel exhausted now. I'm thinking back on my life, how intense this is. But right now, I feel peaceful.

Wolinsky: During the taking on of the dependent identity, worthless, what did you have to pretend that you didn't understand?

Barbara: Well, that I didn't understand what I was feeling?

Wolinsky: Are you still pretending that?

Barbara: No, I can't.

Wolinsky: Recall a forgotten pretend.

Therapeutic Note

People pretend in order to survive. Then they forget they are pretending. This forgotten pretend becomes their persona. Persona comes from the Greek meaning mask.

Barbara: You mean like my second marriage (laughs). I think I pretended in a sense, you use the term, that I didn't know what I was feeling. I had a lot of feelings but I didn't listen to them or thought they were crazy.

Wolinsky: Recall another forgotten pretend.

Barbara: Well, I am smart but I pretend that I'm not.

Wolinsky: Recall another forgotten pretend.

Barbara: I am just getting a flood of different times that I knew something was going on but I told myself that I didn't know it. You know, about people, just hits on people, just stuff.

Wolinsky:	Let all of that filter through and just notice and acknowledge the flood of memories. Recall another forgotten pretend.

Therapeutic Note

When these types of questions are asked, lots of memories arise. It is important to observe and acknowledge the memory. However, it is not important to analyze each one. Ramana Maharishi once said, "When you are cleaning your house, it is not necessary to analyze the dirt."

Barbara:	Now, I'm losing track. I'm pretending that I don't know what I am feeling.
Wolinsky:	Recall another time that you had a pretend but you forgot that you pretended.
Barbara:	I used to do wild things with my sister but it wasn't me. I would pretend I was like her and go out with these weird guys.
Wolinsky:	Recall a time you pretended to be worthless when you really knew that that wasn't so.

Therapeutic Note

Remember the False Core is a concept that you believe (pretend is true). Make the implicit explicit.

Barbara:	Well, see, that fits in under the second marriage. Like knowing these things but acting like I didn't know. Worthless has always been my word. Not valuing myself.
Wolinsky:	Regarding worthless or being of no value, recall another forgotten pretend.

Barbara: When I would get awards in college, I wouldn't go pick them up. It was like devaluing that, I think it was devaluing me somehow. That just came to mind.

Wolinsky: How does worthless seem to you now?

Barbara: It seems multi-purpose. It's like one of those cleaners. It cleans everything.

Wolinsky: Being worthless and having no value has a lot of value?

Barbara: It had a lot of value.

Wolinsky: How are you feeling right now?

Barbara: I feel good. I feel lighter. I feel like making jokes.

Wolinsky: I want to ask you a couple more questions. If worthless was no longer an issue for you, what would you be afraid that you would be responsible for?

Therapeutic Note
People do not take responsibility for their experience. Responsibility is a willingness to realize that you (nervous system) placed an organizing belief (in this case, the False Core) around yourself and have been living out of it.

Barbara: Telling the truth.

Wolinsky: Which is?

Barbara: At any given moment, it could be whatever is true for me.

Wolinsky: What's the truth about you right now?

Barbara: I'm glad I did this process.

Wolinsky: *Prior* to you taking on this concept called worthless, no value, how big was your consciousness?

Barbara: Big, like galaxies.

Wolinsky: And after taking it on, how big was your consciousness?

Barbara: Well, I see this dark little apartment.

Therapeutic Note

This is an important question because it is *prior* to taking on a structure that our consciousness is unlimited. By taking on a structure, it shrinks. This is why Nisargadetta Maharaj was so into *prior to* and saying, "Don't believe anything, question everything."

Wolinsky: How are you doing now?

Barbara: Great—open—spacious.

CONCLUSION

(FOR THE GROUP)

1. Without using your thoughts, memory, emotions, associations or perceptions, are you worthless, worthy, valuable, valueless, or neither?

2. Without using your thoughts, memory, emotions, associations or perceptions, what does worthy, worthless, value or valueless even mean?

3. Without using your thoughts, memory, emotions, associations or perceptions, what does empty or emptiness even mean?

**NOTICE THE
NO-STATE STATE OF
THE NON-VERBAL I AM**
(Volume III)

CHAPTER VIII
THE WAY OF THE NOT DOER

FALSE CORE DRIVER
I Cannot Do, Decide or Act.

FALSE SELF COMPENSATOR
I have to prove I can do, decide or act, by becoming an over-doer or over-achiever.

This particular False Core "I cannot do" or "I have an inability to do or "I do not do" is widely used and justified in Western cultures. As with all False Cores, the False Self Compensators *cannot do* enough to compensate for the False Core of "I cannot *do*." I have a dear friend who has the largest institute in the world in the area of his speciality. Yet, *subjectively*, he feels as though he does not "do." To overcompensate, he becomes an *over-doer*. In Western cultures, this over-doer over-achiever is highly used because it is socially acceptable. It should be noted, as with all False Core-False Selfs, the experience is not objective but subjective; it is a closed system which no feedback can penetrate.

As with all False Cores-False Selfs there is an obsessive-compulsive tendency to try to overcome the underlying unconscious False Core of "I do not or cannot do enough."

This False Core driver has a strong feeling of "I can't do *that*." There is a real subjective *unconscious* and *unacknowledged* feeling of impotence and sometimes even paralysis. The False Core of "an ability to do" or I *cannot* is *not* based on fact; but is on an *unconscious subjective* experience. The False Core driver is a concept.

This particular False Self Compensator tries to handle its not doing with vanity. Vanity is defined here as looking outside yourself through another's eyes, to see how you are doing. In this way, you perform (unknowingly for your parents) and see yourself through the eyes, of another to make sure you're "doing enough or doing it right." Vanity is a way of distracting. It is rather than me being in my present time experience, I am outside of myself seeing how you see me. So, for example, dressing not so I feel comfortable but so that someone will see or think I'm rich or stylish. In this way, you are neither in your body nor even aware of it. I once knew a woman from Canada who complained about back pain yet continued to wear three-inch high heels. I mentioned to her how bad that was for her back and she said, "I like the look."

This False Self Compensator also uses self-deceit which causes individuals to continually exaggerate their abilities (i.e., ability to do and succeed if they are more on the False Self Compensator side) and to become grandiose about all they did and can do. If they are more fused with the False Core driver, then they are more focused and deceive themselves about what they have not done.

For example, a businesswoman came to me for therapy and she used to say, "I can make it happen. I work 24 hours per day." She not only exaggerated her ability to do but also her "working hours." She was more on the False Self Compensator side. Her self-deceit was exaggerating her achievements or her ability to do.

FALSE CORE DISTRACTOR IDENTITIES

This False Core-False Self uses deceit to hide what it has been doing. This False Core-False Self lies and exaggerates to themselves and others about who they are and what they do or don't do. To illustrate this deceit, I once saw a woman for therapy who was from Florida. She had been having a six year love affair outside of her marriage and her husband knew nothing about it. She hadn't had sex with her husband for years which, of course, he was upset about. "I don't want to cause him pain by telling him," she said, "I'm an

Chapter VIII

honest person." She was not only deceiving herself about what she *did do*, but also her husband. She never really acknowledged what she *did do* (or *didn't do*) in her marriage and felt vicitmized by her husband's anger.

In the United States, many therapies and seminars (leaders) resist the False Core "inability to do" by sales pitches which can be full of deceit. For example, they'll tell you, "This cures (*can do*) everything." "This handles (*does*) everything." "This works (*does it*) for everybody." Take a look at some workshop brochures, one I saw said that, "I made $20,000 last year and I've taken this three day workshop and now I make $250,000" or "This workshop *does* everything."

I once discussed this type of deceit in "selling" workshops and seminars with a well-known seminar leader who said, "Well, we are motivating people to take the workshop." Notice how the *reframe* of motivation *justifies the deceit* in how this workshop represents itself.

For Quantum Psychology, if the workshop is overpriced given the standard for the country and claims results which are fast, painless for everyone and guaranteed, they are playing on this False Core-False Self. In other words, it's a *bluff*. (Ichazo calls one of his polarities the *bluffer*. See below.) It is deceitful to *over*state what a workshop can *do*, which is done to over compensate for its *"inability to do"* of the False Core Driver (of the seminar leader or workshop) which is driving the False Self Compensator sales pitch about what "it" (the workshop) *can and cannot do*.

SPIRITUAL PRACTICE DEFENSE: I Am Not the Doer

Spiritual Practice, enlightenment and merging as an *achievement* or through *achievement* or *doing* or that you somehow can *DO* it by *over-doing* Spiritual Practice→Spiritual Success is an **Oxymoron.** Spiritual Success is *No-I, One Substance*, not "*I* did" through *achievement-I* am Enlightened-*you* can be too if you work hard to achieve, *do* more, *do* surrender more, *do* resistance less, (*do*) take up a spiritual practice—or (*do*) pay me more.

"I am or cannot do" can get confused with the spiritual experience called "I am not doing anything" or "I am not the doer." The difference is—in the False Core-False Self you feel pain about the inability to do with no False Core-False Self, you feel neither bad or good because it is a non-issue. I have seen spiritual teachers and students who confuse the two. There is a Sufi story which illustrates this confusion. Mulla Nasudin had a beautiful flower garden. One day a goat got in and began eating the flowers. Nasudin was furious and began beating the goat. Another man was passing by and said, "What are you doing, beating that poor defenseless animal?" Nasudin said, "I am not the doer, God does everything, I do nothing." "Oh, excuse me for saying anything," the passerby said, "I didn't realize you were such a great saint. Let me touch your feet." The passerby looked around and said, "This is a beautiful garden, who made such a beautiful garden?" Nasudin said, *"I did."*

To illustrate further, I knew a man from Texas who got involved with spirituality and "you are not the doer." He suffered greatly because he used this concept to re-enforce "I cannot do anything." He insisted that "anything I do re-enforces the I do." Therefore "I" cannot do anything was the *spiritualized* use of his False Core Driver. Still, he suffered because he needed to take action, but did not. His "teacher" did not see that this philosophy, though correct at the level of **VOID** and **BEYOND** (to be discussed in Volume III), was hurting people who were not at that level. They were placing this philosophy on top of the False Core-False Self, thus adding another layer on top of it as a "spiritual" justification for not doing. His teacher did not see how such a spiritually important understanding was being used by this man to re-enforce and hurt himself.

QUANTUM PSYCHOLOGY PRINCIPLE:

The teacher must "give a teaching" at the level the student is on; otherwise the student can only use the teaching to re-enforce their False Core-False Self.

As with all False Self Compensators used to compensate for the "inability to do," people try to fight their way out of it. They

Chapter VIII

rarely sit in the "inability to do"; but instead try to fight their way out of it *by doing* which in the end only makes them feel worse, "Am I *doing* it right or enough? Maybe I should *do* it more *do* another workshop or see (do) another guru.

FUSIONS

Many times (as will be discussed later) other False Selfs get used in order to resist the inability to do."

Ichazo's Identities
Ichazo calls this over-efficient and their passion is vanity.
Ichazo's I-dentity #1 is a bluffer.
I-dentity #2 is a skillful person.

Skillful is on one side of the I-dentity, bluffer on the other. For example, there was a woman I met in one of my workshops who was extremely intelligent, went to graduate school, had lots of skills, but was also a *major bluff*. She could do everything. There was deceit and bluff in her "ability to *do* (cure) everything."

As mentioned earlier, workshop advertising contains various bluffs such as "I have" or "You can get" or "You can have much more than is possible." I once asked a workshop leader why they lied about their seminar successes and the workshop leader said, "I believe in the placebo [bluff] effect." Another said, "We'll fake it till we make it."

THE IMPOSTER COMPLEX

Earlier I mentioned the imposter complex. Have you met people who say, "I feel like an imposter." This is a red flag for this False Core-False Self. What happens is the way I get *(fill in the blank)* is by performance or *doing* (see vehicles of merger.) To illustrate, let's begin with the shock of the Realization of Separation and I am separate. Now, as a child, Dad is sitting there and all of a sudden he's watching a baseball game. If I sit down next to him and watch the game and cheer when he does, we are merged. Actually "I'm performing or acting but I'm not getting any satisfaction from it. In other words, I'm acting like I am into the baseball game, but I'm really not."

Regarding the impostor complex I have met doctors who have a degree, but they feel subjectively they are "acting" like a doctor, it's a performance, as opposed to, This is actually my intrinsic experience. It's *an act* to cover the inability to do.

THE REASON FOR I-DENTITY— THE SELF IMPOSED REASONS
I cannot do because (*fill in the blank*)
(I am paralyzed, I am bad, nobody likes me, I am empty).

HEALING STYLE—MERGER THROUGH ACHIEVEMENT AND DOING

In this False Core driver, the (mis)labeled **SPACIOUSNESS** of **ESSENCE** is labeled as empty and then as an "inability to do." I must have *done* something bad, that's why I am separate. Therefore, it is better to *not do* (or else something bad will happen). This creates the feeling of being out of balance and self-consciousness. To eliminate this, a False Self Compensator of over-doing emerges (i.e., *I can do anything*).

QUANTUM PSYCHOLOGY'S FALSE CORE:

"The inability to do" False Self Compensator:
Over-doer or super-achiever.

Distractors:
I-dentity #1: out of balance.
I-dentity #2: trying to be balanced.

Chapter VIII

RESISTANCE TO OR SEEKING
(fill in the blank)

What you might want to begin to ask the I-dentity is, What personal quality of life is most important for me to have? This will give you a hint as to what it is seeking or trying to overcompensate for. If your response is love, look at the False Core driver "I am loveless (described later) as a possibility. If the response is achievement, look at the False Core, "I cannot do." Still, it is not as simple as it sounds. There are myriad techniques, strategies and combinations which resist the False Core.

It can take a while to find your False Core Driver. But when it is realized, it is quite a breakthrough to "get" how your entire life is organized around this one concept.

PROTOCOLS FOR DISMANTLING THE FALSE CORE-FALSE SELF

Once again, the following questions should be considered to be mini-meditations. The purpose is to ask the I-dentity a question and notice what, if anything, pops up.

POLARITY #1—I HAVE AN INABILITY TO DO I-DENTITY—I CANNOT DO

QUANTUM PSYCHOLOGY: INABILITY TO DO/OVER-DOING

1. Where in your body do you experience this *"inability-to-do" I-dentity*? Scan your body and notice where in your body the I-dentity is.

2. Who modeled this *"inability-to-do" I-dentity*. Write down the answer until nothing pops up.

3. When this *I*-dentity emerges and you take it on (identify with) this *"inability-to-do" I-dentity*, is there a feeling of collapse somewhere in your physical body? Write down the answer until nothing pops up.

Therapeutic Note

At this point it is advantageous to externalize the *I*-dentity. This can be done in the following way:

>   Step I: Notice where in the present time body the I-dentity is located.
>   Step II: Notice the I-dentity's size and shape.
>   Step III: Take the label off of the I-dentity. Have it as energy.
>   Step IV: Allow the I-dentity as energy to move from the present time body to another physical location in the room, and then to become solid again.

4. Is this *"inability-to-do" I-dentity*, or does this "inability-to-do" I-dentity feel like "a child" or not having the skill level that was demanded? Ask the *I*-dentity this question and write down the answer until nothing pops up.

5. Has this *"inability-to-do" I-dentity* ever felt stuck because it didn't have the skills that living demanded? Ask the *I*-dentity this question and write down the answer and examples until nothing pops up.

6. When you identify with this *"inability-to-do" I-dentity*, how old do you feel? Write down the answer until nothing pops up.

7. Ask the *I*-dentity to "recall a time in its practicing period (age 2 to 4) when it was learning to walk, to manage food at the kitchen table, to sit up straight at the table, etc., required (possibly too early) to do tasks that it felt it was not ready to do and, hence, it

Chapter VIII

felt an *"inability-to-do"*. Ask the *I*-dentity this question and write down the answer until nothing pops up.

8. What were the expectations of your parents that this *"inability to do" I-dentity* found difficult to meet? (Notice if the *I*-dentity gets or trance-fers these demands onto others, like teachers or authority figures.) Ask the I-dentity this question and write down the answer until nothing pops up.

9. During the emergence→identification of this *inability-to-do" I-dentity*, was there any chaos that occurred? Write down the answer until nothing pops up.

10. During the emergence and taking-on of the *"inability-to-do" I-dentity*, was there any level of expectation or expectations from another that caused it to try to do things that you were not ready to do? Ask the I-dentity this question and write down the answer until nothing pops up.

11. Notice in your present-time life the feeling state of discomfort when living demands that you do things that you are not ready to do, (i.e., in a relationship—possibly to make a commitment—and changes that you are not ready to make in living situations (like moving, or breaking up a relationship; things that you are not ready to experience, like the death of a partner)) and notice if *"I just can't do"* comes up. Ask the I-dentity this question and write down the answer until nothing pops up.

12. Notice that the *"inability-to-do" I-dentity* is age-regressed (stuck in time).

13. During the time of taking on this *"inability-to-do" I-dentity*, what decisions did you make about living? Ask the *I*-dentity this question and write down the answer until nothing pops up.

14. Ask the *"inability-to-do" I-dentity*, what catastrophic fantasies it creates. Ask the *I*-dentity this question and write down the answer until nothing pops up.

15. During the process of taking on this *"inability-to-do" I-dentity*, notice what expectations, demands or criticism came from others that now have been internalized as a critic, demander or internalized voice that demands that you take action. Ask the *I*-dentity this question and write down the answer until nothing pops up.

16. During the process of taking on this *"inability-to-do" I-dentity*, what expectations appear as though they are coming from oneself which were really internalized from others? Ask the *I*-dentity this question and write down the answer until nothing pops up.

17. Notice where in your body "this *"inability-to-do" I-dentity* is now. Write down the answer.

18. Peel it back and go into the **SPACIOUSNESS** of your **ESSENCE** prior to the *"inability-to-do" I-dentity* or the (mis)label of emptiness.

19. From inside that **SPACIOUSNESS**, ask the *"inability-to-do" I-dentity* and the empty label, "What are you seeking more than anything else in the world?" Allow the *I*-dentities and labels to answer.

20. Experience that sought, essential quality in the **SPACIOUSNESS**.

21. From "back there," in the **SPACIOUSNESS** feeling that essential quality, notice the size and shape of the *"inability-to-do" I-dentity* and the empty label. Notice the space that surrounds them.

22. See the *"inability-to-do" I-dentity*, the empty label and the **SPACIOUSNESS** that surrounds them as made of the same substance, as you continue to feel the essential quality.

FALSE SELF COMPENSATOR: THE OVER-DOER OR OVER-ACHIEVER I-DENTITY

1. Notice where in your body you experience the "over-doer, over-achiever" I-dentity. Ask the *I*-dentity this question and write down the answer until nothing pops up.

2. Stand up, take-on and exaggerate the "over-doer, overachiever" I-dentity's posture. Now, stand up and exaggerate the "inability-to-do" I-dentity's posture.

3. Step outside of both of them and observe the two postures: "overachiever, over-doer" and the "inability-to-do" I-dentity.

4. Take a few minutes to go back and forth from one to the other, so that you can be aware of the False Core ("inability to do") and the False Self ("over-doer" achiever).

5. Who modeled this *"over-doer, overachiever" I-dentity*? Write down the answer until nothing pops up.

6. Has this *"overachiever" I-dentity* ever been used to get love? Ask the *I*-dentity this question and write down the answer until nothing pops up.

7. Has this *"overachiever" I-dentity* ever been used as a way of substituting achievement for love? Ask the *I*-dentity this question and write down the answer until nothing pops up.

8. Has the *"over-doer, achiever" I-dentity* ever been used to control another's feelings toward you? Ask the *I*-dentity this question and write down the answer until nothing pops up.

9. Has the *"overachieving" I-dentity* ever been used as a way of impressing others? Ask the *I*-dentity this question and write down the answer until nothing pops up.

10. Has the *"overachieving" I-dentity* ever created fantasies as a way of resisting the False Core of an "inability-to-do?" Ask the *I*-dentity this question and write down the answer until nothing pops up.

11. Ask the *"Inability to do/achieve" I-dentity* to recall several times that it felt impotent and an "inability-to-do," and it created fantasies to combat and overcompensate for these feelings. Ask the *I*-dentity this question and write down the answer until nothing pops up.

12. Has this *"inability to do" I-dentity* ever felt that by using the *"overachiever" I-dentity*, even after an achievement, that there was actually no internal sense of achievement? Ask the *I*-dentity this question and write down the answer until nothing pops up.

13. Ask the *"inability to do" False Self* to recall a time when it felt no matter how well it performed, it would never be enough. Ask the *I*-dentity this question and write down the answer until nothing pops up.

14. Ask the *"inability to do" I-dentity* to recall a time that it felt it used this "over achieve" I-dentity as a way of deceiving others, and hence, felt like an imposter. Ask the *I*-dentity this question and write down the answer until nothing pops up.

15. Has the *"inability to do" I-dentity* ever been used to be too busy to let people have contact with you? Ask the *I*-dentity this question and write down the answer until nothing pops up.

16. During the emergence→I-dentification of this *"over achieving" I-dentity* that is into performance, was there any chaos

Chapter VIII

that occurred? Ask the *I*-dentity this question and write down the answer until nothing pops up.

17. Ask the *"inability to do" I-dentity* to recall a time in its past, or in its first few years of life, where it had to perform (for its parents) so that it could feel as though it had an individual self. Ask the *I*-dentity this question and write down the answer until nothing pops up.

18. Ask the *"inability to do" I-dentity* to recall a time in the first few years where you had to perform for your parents in order to get love. Ask the *I*-dentity this question and write down the answer until nothing pops up.

19. During the process of I-dentification with this *"over achieving" I-dentity*, were there any decisions, beliefs, or assumptions that were made? Ask the *I*-dentity this question and write down the answer until nothing pops up.

20. During the emergence→I-dentification of this *"over achieving" I-dentity*, were there any fantasies that were created—(for example, receiving praise in the future, winning the Nobel Prize, winning an Oscar, winning some kind of trophy or award)? Ask the *I*-dentity this question and write down the answeruntil nothing pops up.

21. Notice where in your body you experience the *"over-achieving" I-dentity* and the empty label. Write down the answer until nothing pops up.

22. Peel it back and go into the **SPACIOUSNESS** within the body.

23. From inside the **SPACIOUSNESS**, ask the I-dentity and the empty label, "What are you seeking more than anything else in the world?" Ask them this, allow them to answer and write down the answer.

24. Experience that quality inside the **SPACIOUSNESS** of your own **ESSENCE**.

25. Notice the size and shape of the I-dentity and the empty (label). Notice the space that they are surrounded by or floating in. Experience them as the same substance. Let them dissolve as you experience the essential quality inside your body.

Demonstration

Sylvia is a 50 year old woman with four children who lives in Ohio. We are going right into her False Core driver.

Wolinsky: Who modeled this "inability to do" I-dentity?

Sylvia: My father. He did for others and had numerous things that he always wanted to do but he could never get past taking care of everybody else or doing what my mother wanted to do.

Wolinsky: What are you feeling right now?

Sylvia: I am going numb. I'm getting reasonable. I'm telling you his story.

Therapeutic Note
Here we see the (inability to do) going to the overly reasonable False Self (see next chapter).

Wolinsky: So create the feeling of numb and then go into your head and become reasonable.

Therapeutic Note
Do what you do knowingly, consciously, intentionally with awareness.

Sylvia: I'm in it.

Wolinsky: Do it intentionally.

Sylvia: It's like a pulling up in my head. It's doing a sort of performance. I am able to tell you his story.

Therapeutic Note
The "inability to do" False Core driver can easily go into the "as if" and begin acting and performing.

Wolinsky: Have you ever taken on this "inability to do" I-dentity because there was an external demand of skills which you didn't have as a child?

Therapeutic Note
The "inability to do" often goes into the False Self compensator by becoming an over-achiever. This occurs often times when parents ask children to *act* older than they are.

Sylvia: Yes. I was the oldest of three children. I would be left to clean the entire house. We didn't have a big house. My mother would go off and do errands.

Wolinsky: So what did you feel when this demanding force came at you.

Sylvia: I would get limp. I would be unable to do it.

Wolinsky: Create this force coming at you, which is demanding you to do something that you can't do. (See Chapter XVIII on Force Theory) for an explanation.)

Sylvia: Yeah, I feel it.

Wolinsky: Tell me something you decided about this situation.

Sylvia: If somebody tells me to do something, I will do or pretend to do it. I will dissociate from it because I don't want to do it.

Wolinsky: This force came at you called "demand" and you went limp, what did you decide about yourself?

Sylvia: To not do it.

Wolinsky: What did you decide about another?

Sylvia: That they were the boss.

Wolinsky: What did you create in response to them being the boss?

Sylvia: In the end doing it. Going limp. Not to do it. To want to do something else instead.

Wolinsky: Tell me something you decided about yourself from that place of limpness.

Sylvia: I will force myself to *do* it.

Therapeutic Note
Here is the False Self compensation or over (false) doing

Wolinsky: Tell me something else you decided about yourself.

Sylvia: That I don't want to do it.

Wolinsky: Tell me something else you decided about yourself.

Chapter VIII

Sylvia: If I got away from them, I would be able to do.

Wolinsky: Tell me something else you decided.

Sylvia: I was worthless for wanting to do that because it wasn't doing or giving for someone else.

Therapeutic Note
Here the "inability to do" moves to both the False Core compensator of over-doer and the over-giver to get worth or value.

Wolinsky: Tell me something else you decided.

Sylvia: That I was inadequate and otherwise I would be able to do it. Somehow, it must be I am inadequate. It must be, something is wrong with me. The old saying, What's wrong with you.

Therapeutic Note
She moves from "the inability to do" to "I am inadequate" to "there must be something wrong with me." Do not get distracted as she distracts herself, stay and keep her on "the inability to do." No matter what someone presents, stay on the False Core. They will try to distract you or themselves from the False Core. "Don't let them!"

(continued)

Sylvia: It's more weight.

Wolinsky: Notice the weight, the size, the shape, Yes. Notice the space that is floating.

Sylvia: It's on me.

Wolinsky: Yes. Now, take the label off it and have that which is on you, have that as just energy.

Sylvia: I feel like I am shaking, a strong vibration.

Wolinsky: Feel the shaking and the vibration.

Sylvia: It feels sort of like I could go crazy.

Wolinsky: Feel that.

Sylvia: I a-void this. I have a crazy brother. He went crazy. I can't be crazy.

Wolinsky: You have a crazy brother. What does that mean?

Sylvia: A typical psychosis.

Wolinsky: What I would like for you to do is feel that crazy, look at me.

Sylvia: I shut it off. I don't want you to know that.

Wolinsky: Say that to me. "I don't want you to know that I stop myself from going crazy."

Sylvia: I don't want you to know that I stop myself from going crazy.

Wolinsky: Say to me, "I really resist *doing* crazy."

Sylvia: I really resist *doing* crazy, crazy like my brother, crazy like my family. I guess I do resist going crazy.

Wolinsky: Define crazy.

Sylvia: I'm losing my ability to understand what it means.

Wolinsky: Define crazy for me. What does that mean?

Chapter VIII

Sylvia: If somebody is crazy, they have no ability to be in the world. *They can't do anything.*

Wolinsky: Have an image of your brother over there (another location in the room).

Sylvia: Okay.

Wolinsky: Where in your body do you feel your crazy brother?

Sylvia: In my pelvis.

Wolinsky: Take the label off and allow the "crazy" to go back to your brother as energy.

Sylvia: Okay.

Wolinsky: How do you feel?

Sylvia: Clearer. I feel proud of myself. Because I could do it and understand you.

Wolinsky: You did do that.

Sylvia: I did do that, Yes.

Wolinsky: What happened?

Sylvia: Seems like such a baby step, seems like such a baby thing to do. I'm proud of myself for doing it. I mean 50 years of this.

Wolinsky: Where in your body do you feel the little girl I-dentity who has an inability to do?

Therapeutic Note

Notice *it is not her*. It is a little girl I-dentity.

Sylvia: It starts right here and it goes right up the middle.

Wolinsky: So have an image of the little girl and put it over there (in another physical location).

Sylvia: Okay.

Wolinsky: How do you feel looking at it now?

Sylvia: It feels/looks small.

Wolinsky: How do you feel?

Sylvia: It's there, I'm here, and I feel fine.

Wolinsky: Now, the craziness.

Sylvia: Yeah,, I feel the craziness cellularly.

Wolinsky: Cellularly?

Sylvia: The inability to move is cellular.

Wolinsky: Now, put that inability to move over there (another part of the room). Where in your body do you feel the crazy? Now, take the crazy and put the crazy there (another part of the room).

Sylvia: Okay.

Wolinsky: How do you feel looking at this?

Sylvia: That it makes sense.

Chapter VIII

Wolinsky: What's your experience like right now?

Sylvia: I feel calm.

Therapeutic Note
Once the client dis-identifies and externalizes from the I-dentity, they let go.

Wolinsky: So you have this "inability to do" and there we have the crazy family situation. How does *doing* seem to you now?

Sylvia: I hate myself.

Wolinsky: Make this statement to your family. I hate you, I don't hate me.

Therapeutic Note
ESSENCE-I AM never hates itself. She has swallowed or retroflected the hate for her family to herself. In other words, since I could not hate you, I will put it back on myself and hate myself.

Sylvia: (To her family) I hate you, I don't hate me. I hate me too.

Wolinsky: Look into your mother's eyes. Into your father's eyes. Say I hate you, I don't hate me.

Sylvia: I hate my mother. I hate you, I don't hate me. I hate her. She hates me right back because I'm hating her back. She hated me first. I will hate her back.

Wolinsky: Say that to her, I hate you because you started it.

Sylvia: I hate you because you started it because you hated me first. That's the truth. That's the God's honest truth.

Wolinsky: I want you to say to your mother, "Since I couldn't tell you I hate you directly, I hated myself."

Sylvia: Since I couldn't hate you directly, I hated myself.

Wolinsky: Notice that behind the image of your mother there is a lineage of your mother's mother, and your mother's mother's mother.

Therapeutic Note
False Core driver's have a lineage.

Sylvia: Well, I know at least three generations. Yes.

Wolinsky: So see your mother and grandmother, great-grandmother and how ever far back it goes.

Sylvia: They all hated their mother. I mean the mother hated the child and the child hated the mother.

Wolinsky: I want you to notice the energy in this lineage.

Sylvia: I'm in line to it.

Wolinsky: I want you to take the label off of "it" (on you) and reverse the energy flow. In other words, it (lineage energy) is coming at you. Reverse it (the lineage energy) and let it go back to them.

Sylvia: It's like I want to protect my daughters.

Wolinsky: Yes.

Chapter VIII

Sylvia: And I always feel like they are going to hate me. I always assumed that they were going to hate me no matter what I do.

Wolinsky: Now, I want you to make a statement to your mother and to that lineage, but to your mother more directly. "Since you hated everything I did, *I stopped doing.*"

Sylvia: "Since you hated everything I did, *I stopped doing.*" "That's true."

Wolinsky: Now, allow the little girl I-dentity to say to mom, "I froze my hatred."

Sylvia: (Through little girl I-dentity) "I froze my hatred."

Wolinsky: "I froze my inability to do."

Sylvia: *My hatred is my inability to do.* My frozen hatred is my inability to do just by not being able to say to her, I hate you what you have done to me. I can't do anything else. I haven't been able to do anything else.

Wolinsky: Say that again.

Sylvia: Just by being frozen in my hatred and not being able to say "I hate you" and to acknowledge your hatred for me, *I have been unable to do anything else my whole life. I have been paralyzed by hatred.*

Wolinsky: Say that to your mother.

Sylvia: *I have been paralyzed by your hatred.*

Wolinsky: That the lineage of energetic hatred has paralyzed me.

Sylvia: The lineage of energetic hatred has paralyzed me. All of a sudden all these levels of paralysis are coming up, sexual, thinking. The inability to do is all criticism, hatred, rejection of that.

Wolinsky: How are you feeling right now?

Sylvia: A little bit enlightened. And I never knew that it was still impacting me. I moved to Ohio. I figured if I was around a lot of people who were blond and blue eyed that didn't look like me…

Wolinsky: If you got away from them, you would be (*fill in the blank*).

Sylvia: I thought I was running to freedom just to be me and I was sort of doing the thing where I was carrying the Catholic cross of the culture with me. I just took it and set it up in my household.

Wolinsky: How does this lineage seem to you now?

Sylvia: I understand what I thought I understood before. I understand it. I have a deeper sense of the restriction.

Wolinsky: I want to know where in your physical body you feel the hatred.

Sylvia: Right now, I don't feel it. It is over there (with mom).

Wolinsky: Okay. How are you feeling right now?

Sylvia:	Lighter. I don't feel afraid. I was afraid to say I guess what I said. I didn't realize I had laid down and die for a long time. I did do it. I did what I thought I wasn't doing. I died. (Crying) I never knew that. I thought I was not dying.
Wolinsky:	Now, through this little girl I-dentity, make this statement to mom, "Since I couldn't kill you, I killed me."
Sylvia:	Since I couldn't kill you, I killed me.
Wolinsky:	Say that again. Since I couldn't kill you, I killed me.
Sylvia:	I think she tried to kill me once. I think she did try to kill me once when I was a baby. Since I couldn't kill you, I tightened up and killed me.
Wolinsky:	What I am going to ask you to do is keep your eyes closed for minute and imagine your mother is directly in front of you. I am going to be behind you just to lift your hands. Just imagine that you are going to strangle your mother. (I have her lift her hands toward an image of mom. I am asking her to strangle mom.)
Sylvia:	I can't do it Stephen.
Wolinsky:	Okay, then instead strangle yourself.
Sylvia:	(She moves her hands to strangle the image of mom and then she strangles herself) That feels better. I've done this before. There is such a resistance, I can't do.

Wolinsky:	So move like this (as to strangle your mother) and then instead strangle yourself.
Sylvia:	Yeah, it's amazing. I would love to be able to do (strangle my mother) but my hands won't go.
Wolinsky:	Yeah, they go like that. (I place her hands on her own throat so she can strangle herself.) Does that feel more comfortable?
Sylvia:	Yeah.
Wolinsky:	So answer the questions from here (strangling herself - *her* hands on *her own* throat) Say to your mother, "Since I couldn't *strangle you, I strangled me.*"
Sylvia:	Since I couldn't strangle you, I strangled me.
Wolinsky:	You look much more comfortable.
Sylvia:	I feel very comfortable. Honest to God I do.
Wolinsky:	*Now, I want you to notice the lineage of strangled women.*
Sylvia:	It's the truth (crying).
Wolinsky:	Take a breath.
Sylvia:	(Crying)
Wolinsky:	How does it feel to be the holder of that lineage?
Sylvia:	That explains my not wanting to get near men who sort of look like me (She is dark haired, dark eyed. Her mother and father are southern European with

dark hair and dark eyes,) because I've had babies who look like me so I had 6 foot tall, fair-haired, blue eyed men to make my babies with. My daughters, people used to think I was the hired servant because my babies were blond and blue-eyed. It makes sense. So my daughters wouldn't be like me. They wouldn't look like this. (A strangled woman) They wouldn't have to be like me. (Crying intensely)

Wolinsky: So you were trying to break the lineage in your own way?

Sylvia: Yes. I did that.

Wolinsky: *You did do that. You did* try to break the lineage that way.

Sylvia: I was trying to do that without knowing it.

Wolinsky: So right now are you willing to knowingly begin to break that lineage?

Sylvia: Knowingly, Yes. I'm shaking inside though.

Wolinsky: It's whole generations.

Sylvia: I'm holding on.

Wolinsky: Is your grandmother dead?

Sylvia: Yes.

Wolinsky: She can't hurt you. Is your mother dead?

Sylvia: No.

Wolinsky: She still can't hurt you.

Sylvia: I know. I moved away from her.

Wolinsky: She can't hurt you anyway.

Sylvia: I know but I can hurt my daughters.

Wolinsky: Not if you don't pass on the lineage.

Sylvia: I try not to, but I may have without knowing what I haven't known so far.

Wolinsky: For your daughters' sake, are you willing to break this lineage?

Sylvia: Oh, my God, Yes. Yes.

Wolinsky: What are you feeling right now?

Sylvia: Relief. I feel stronger. I feel like for me it has wasted 50 years but, Jesus, it could have taken a lot longer.

Wolinsky: So, for your children you are willing to break the lineage. Now, for yourself are you willing to break the lineage?

Therapeutic Note

This is a family therapy approach (i.e., children as therapeutic leverage). People won't do it for themselves. They will do it for their children.

Wolinsky: How are you doing right now?

Chapter VIII

Sylvia: I'm shaking physically. I feel disorganized, rearranged. My body doesn't feel like it's all pieces or in the same place. I don't have a sense of being arranged the way I was when I came here so I'm kind of shaking. What I am feeling is all this jumbleness. Like I am full of these tiny little ping pong balls and they are all just jumbling around in a chaos kind of feeling. In my legs particularly where I hold my stress. Where I used to feel like I had to hold my ground. I'm feeling as if I am transparent. You could see all these different colored little tiny balls just jumping around. It feels like that. So I am not quite sure where my skin is any more and I am usually used to that boundary of knowing exactly where my skin is. It feels like I don't have a body but I feel awake. I feel like I don't have a body. It's not like a body I have had anyway.

Therapeutic Note
When you begin to drop the False Core, the nervous system reorganizes and "you" are no-body. Thus, I Am No-body, yet awake, is the wakeful sleep of mediation.

Wolinsky: Okay. If we have the little girl I-dentity over there who feels an "*inability to do*" and all this hatred and so on. Here's this lineage there which goes all the way back through here. How do they seem to you now?

Sylvia: I feel really sorry for them. I just feel like Oh, my God. All the wasted lives. It is like a holocaust. And not just for me but to my mother, my grandmother, but from me out, the holocaust. The hatred sort of dissipates in a way. You know it is like, How could you hate me? That's the root of it. I mean it is not sadness. I don't feel grieving.

It's a holocaust. All these women died or killed themselves, were the walking, living dead.

Wolinsky: Wearing black?

Sylvia: Yeah, and that struck me when I visit Greece. The Babushka thing, the veiling and grieving of self.

Wolinsky: So now when you feel the chaos in the cellular way. So what I would like for you to do is to focus on that chaos as energy.

Sylvia: It feels like it. That is what it does feel like.

Therapeutic Note

I have a woman from the group sitting with her back to the client. I have Sylvia put a black scarf (Babushka) on her head. I am asking Sylvia to take it off and to put it on the woman's head.

Wolinsky: Notice just as we go along how it reorganizes. Now, I want you to see, as you look at the back of her head and the lineage beyond her, that you can't change that lineage. You can't change this part, this mother. Are you willing to give the lineage back to its original source? Energetically do that?

Sylvia: All that's true. I know I can't change the past. Why instead of not giving it back, can I just burn it.

Wolinsky: Because it doesn't work and it is not yours, and I would like for you to give the energy back. You don't have to. Feel the stuckness and reluctance to give the energy back to the lineage.

Sylvia: That's true. To get punishment, to get back love. To get repressed, you support them and take care of them. You get negated. You take care of their needs.

Wolinsky: I want you to hold that there and I want you to put all of the energy of that. You get punishment, you have to give back love. You get hatred, you have to give back love—I want you to put that in your body.

Sylvia: I feel more rigidness.

Wolinsky: Well, you are holding the energy of the lineage and putting it inside you.

Sylvia: Well, how do I put it in me? I feel stiff. It feels like this energy of stuffiness. I rigidify. I feel very rigid. I can feel nothing. I mean if this was just energy, I can't relax.

Wolinsky: So, as you hold onto this (the black scarf - Babushka), is there an illusion that if you hold onto the lineage, if you hold on to it, then it doesn't have to be passed on.

Sylvia: I think so somehow. Yeah, it's like having a baby out here, out between my legs and holding onto this up here. They get free of it. I just got that image. The baby popping out but I kept (the scarf, Babushka) this up here separate from the baby. The baby is free that way.

Wolinsky: So the only person that dies through this is you.

Sylvia: Well, Yeah, by wasting my time by holding onto this. I certainly don't get a lot. But it is like, you've got the poison that if the little thing drops and

| | breaks and the whole world will die, then certainly it is an important job, isn't it so that life goes on? |

Wolinsky: *So you did do something?*

Sylvia: Yeah, I guess I did. It is like holding on. It is like, don't let it hit the ground and break because people will die.

Wolinsky: Can you say, so I will die instead?

Sylvia: When I say this and that image comes, it doesn't feel like dying. It feels like protecting. It feels like taking care of.

Wolinsky: In the process of taking care of, do you have to kill yourself?

Sylvia: No, just give up a lot.

Wolinsky: Give up yourself.

Sylvia: Yeah, my time, right. Or doing, because I can't do anything.

Wolinsky: I understand. Say that again.

Sylvia: *I have to give up doing because I've got to hold onto this (the lineage). I can't do anything.*

Wolinsky: So as long as you are holding onto this (the lineage), you can't do anything.

Sylvia: That's true.

Wolinsky: Say that again.

Chapter VIII

Sylvia:	*As long as I hold onto this, I can't do anything. I'm not free to do anything.*
Wolinsky:	Tell me something else in this lineage the little girl's I-dentity is holding onto.
Sylvia:	Well, it came to me outside. I went out East in the fall. I was hanging around my mother, helping her out. And so I was going to all of these places where people look like me. After being out West where most people are blond and blue eyed, I had this sense of belonging. These women that looked like me. I felt like I belonged.
Wolinsky:	I want you to make a statement to me. By focusing on the women and shunning from the men, I belong.
Sylvia:	The dark women like me. The dark haired, dark skinned women. I felt like I belonged. I felt comfortable.
Wolinsky:	And shunning from the men.
Sylvia:	Yeah, the men.
Wolinsky:	Is that part of the lineage?
Sylvia:	Oh, absolutely. Men like them in particular. Men like them and me make more babies like me. I don't like men like them.
Wolinsky:	Notice where in your present time body you feel this and take the label off and allow the energy to go back through the little girl I-dentity and then back to the lineage.

Therapeutic Note

The "inability to do" False Core driver and all its ideas are part of an age-regressed little girl I-dentity. These ideas are attached to her present time body. In this way, she cannot experience present time as present time. Rather, she experiences present time through the little girl I-dentity. Taking the label off of her present time body it goes back to the little girl, then to lineage. This is the original roots of "her" present time ideas about herself and life.

MAKE THE IMPLICIT–EXPLICIT

Wolinsky: Tell me something else in this lineage that you are holding onto.

Sylvia: I realize there is a whole set of family rules.

Wolinsky: Tell me a rule.

Sylvia: You always pretty much have to be poor even when you make money. It's better that you are poor but if you do sort of make it, you put it away so nobody knows you have it. So even you don't know you have it. You leave it to your children. Then the children feel guilty that you died without spending it so they won't spend it. This is true.

Therapeutic Note

Keep repeating the process about 1) where in the body; 2) take the label off; 3) allow the energy to go from her through the little girl I-dentity to mom and the lineage.

Wolinsky: Tell me another rule, a lineage rule.

Sylvia: The oldest daughter takes care of the mother, the grandmother and everybody else in the family. The oldest daughter. Sons don't have to do that.

Chapter VIII

Wolinsky: Tell me another rule in the lineage.

Sylvia: If you are a woman in that culture, in that lineage, you are not to be successful. You are not to have an occupation other than to be mother and daughter or granddaughter.

Wolinsky: Tell me another rule of the lineage.

Sylvia: Mostly you don't go to school.

Wolinsky: Tell me another rule of the lineage.

Sylvia: You have to feel guilty all the time. You are lucky to be alive with all the misery in the world, everybody is dying, everybody is sick. Don't think positive because at any moment, that could change.

Wolinsky: Tell me another rule of the lineage.

Sylvia: For women, it is to be Holy Mary, the Virgin, or otherwise you are Mary Magdalene. So sexually, it's repression. You don't talk about, and women don't talk to each other about real things like sex, men, feelings, postpartum depression, nothing. What's wrong with you, take care of those babies, you are lucky you have those babies. You are lucky to be alive or you are lucky they've got two arms and two legs.

Wolinsky: Tell me another rule of the lineage.

Sylvia: Have no needs.

Wolinsky: Tell me another rule of the lineage.

Sylvia:	Be stupid and don't even know what needs are. My mother would be real smart at work. She would come home and she would be stupid. My sister did the same thing. So I didn't go to work and just stayed stupid because it was just too much for me to have to split apart. Keep you mouth shut. Take care of everybody.
Wolinsky:	Tell me another rule.
Sylvia:	It's better to suffer then to never live it all.
Wolinsky:	Tell me another rule. Tell me something a psychic Babushka contains.
Sylvia:	You wear it even out in the hot sun, you know. So it is sort of contains all of the limits.
Wolinsky:	Tell me the limit a psychic Babushka contains.
Sylvia:	You kind of feel hot and want to take it off even if it is hot and sunny out.
Wolinsky:	Tell me something else the psychic Babushka contains.
Sylvia:	I put my head down and what I'm feeling is, it contains the church and it keeps it on you.
Wolinsky:	Tell me something else the psychic Babushka contains.
Sylvia:	It contains anything you think, anything you feel, anything you want to do, anything you want to say, anything you want. Certainly it keeps you blind from anything you need. What is a need?

Chapter VIII

	It's also a gag order from God for women. It is. I feel mad. I'm angry.
Wolinsky:	So where in your body did you hide those forces?
Sylvia:	My ovaries. That's why I have four kids. They could do it for me. They could be those things.
Wolinsky:	So you feel it in your ovaries?
Sylvia:	All of a sudden the sense of two ovaries pop right up.
Wolinsky:	So I want you to go in and feel that and rather than it going out of you through your vagina, I want you to take the label off of all that very concentrated energy.
Sylvia:	There is a sense of not safe up here on my shoulders.
Wolinsky:	How far up does it go before it collapses around not safe?
Sylvia:	No further than this. Just to my navel. I'm just aware of how my shoulders feel. They feel very weighted. It feels sort of like a protection.
Wolinsky:	How does the Babushka seem to you now?
Sylvia:	It's like shackles and chains. I am willing to give it away. I'm willing to put it over there.
Wolinsky:	Well, if you want to do that, I want you to do that very slowly. What are you feeling right now?

The Way of the Human • The False Core and the False Self

Sylvia: I have a headache. I have an awareness of this weight on my shoulder and I just feel tired of it.

Wolinsky: How do your ovaries feel?

Sylvia: I didn't check them. I'm just more aware of this. Small but a little bubbly.

Wolinsky: So if you do this, I want you to do this very, very slowly with awareness from your ovaries so that the energy of your ovaries is coming through you and taking that off and putting that on your mother which goes back to the lineage.

Sylvia: Do it from my ovaries?

Wolinsky: Yeah, if you can't, that's not a problem.

Sylvia: Well, I guess I don't know It's from my ovaries. But his feels cramped too. It's uncomfortable. I just have this desire now to put it back. If that's coming from my ovaries, Well—

Wolinsky: How does your ovaries feel as the desire builds?

Sylvia: They are aching. There are cramps there too.

Wolinsky: So very slowly, stand up. How does the energy in your ovaries feel?

Sylvia: Cramped. I feel cramps.

Wolinsky: Now, take the label off of the cramp and have that as energy.

Sylvia: There is this kind of squeezing. Like it is all cramped.

Wolinsky:	Take the label off all of that, and have that as energy. Let me know when that energy begins to spread so that it moves up through your torso and then shoulders, head and face. Through your hands and fingers. Take the label off of rigidity. Let me know when it goes through your face, hands, down your arms.
Sylvia:	It feels like sort of, if I do that, then both arms go down. I just want to drop it. If I want to do this, I've got to keep—
Wolinsky:	I don't want you to do it from push (i.e., "I have to do.")
Sylvia:	I cannot do it yet.
Wolinsky:	I don't know if you are ready yet, this feels like an O.K. place to stop for now.
Sylvia:	Yes—I'm O.K.

Therapeutic Note

People take on my mother's False Core to have a relationship with her. Children learn that, and that is the only way to have a relationship with mom, because they've lost their **ESSENCE**. So now the only way to have a relationship with mother is to take on her False Core. In Quantum Psychology, we want to be able to have an **ESSENCE to ESSENCE** relationship without taking on her False Core. So rather than psychology having either you are totally separate and individuated, or on spiritual trips, you are only totally all universally connected and no individuality. In multi-dimensional awareness, you could have individuality in the context of unity. So the context is always **ESSENCE** and beyond while you have separation.

CONCLUSION

(TO THE GROUP, EYES CLOSED)

1. Without using your thoughts, memory, emotions, associations or perceptions, are you a doer or not doer or neither?
2. Without using your thoughts, memory, emotions, associations or perceptions, what does doing or not doing even mean?
3. Without using your thoughts, memory, emotions, associations or perceptions what does empty mean.

**NOTICE THE NO-STATE
STATE OF THE NON VERBAL I AM**

CHAPTER IX
THE WAY OF THE INADEQUATE

FALSE CORE DRIVER:
I am inadequate.

FALSE SELF COMPENSATOR:
I must prove I am not inadequate and I must prove that I am adequate and smart.

The false conclusion of this False Core Driver is "I am inadequate." As with all the other False Core drivers, "I am Inadequate" drives the machine of its psychology and associations.

In review, a False Core Drives the False Self Compensator to obsessively-compulsively reach Nirvana (merger with mom, later psychologized as "health" or *spiritualized* as God) through becoming or proving adequacy. Remember in the end, this obsessive compulsive tendency of the False Self to try to prove adequacy is weaker than the obsessive-compulsive tendency of the False Core to prove itself (i.e., in this case, that it is inadequate). Why? Because 1) the False Core driver is the first conclusion and, therefore, organizes and drives the False Self solution like a machine. In short, no False Core Driver—no False Self Compensator. 2) The False Core-False Self is holographic and, as with all False Core drivers and False Core Compensators, it is a closed loop. 3) The False Core ultimately leaves each False Self Compensator exacerbated and frustrated because you can never overcome, heal or transform (a false conclusion) through an over-compensation, because it is a solution based on a false conclusion.

The tragedy of the False Core-False Self-Observer complex is that psychological and spiritual attempts to heal, transform, convert vices (sins) to virtues, move from unhealthy to healthy, disintegrative to integrative, reframe, reassociate, re-decide, get rid of, go beyond, surrender, overcome, etc., etc., etc., is the False Self Compensator doing the action of psycho-spiritual practices—thus, ultimately, these actions of the False Self only serve to re-enforce the False Core.

The False Core driver is simply a false conclusion, a false premise, a false assumption. The False Core driver is misinformation about oneself and must be discarded. The False Self compensator must fail because it is a solution based on a false conclusion.

Chapter IX

The False Self Compensator is doomed to failure because it defines the solution from a false conclusion (False Core).

The False Core is only a concept. The False Core needs to be seen as just that and not to be flirted with as in trying to heal or transform it or make it better, etc.

The False Core-False Self occurred out of the "mind" of an infant (5-12 months). Quantum Psychology sees the "way out" by the "way in":

1. Being aware of your False Core driver.
2. Observing your False Core.
3. Being willing to feel the pain of your False Core driver.
4. Tracing the "chains" of associations and all behavior, etc. back from the False Self Compensator to your False Core driver, thus breaking the associational chain and network.
5. Tracing it back to the verbal **I AM** (see Volume III, Chapter The **I AM**).
6. Going into the non-verbal **I AM** (no thoughts, memory, emotions, associations or perceptions, *prior* to the False Core Driver and its verbal formulations and representations.
7. Dismantling your False Core Driver and False Self Compensator.

Pyscho-spiritual Defenses:
1) Proving adequacy through the psychological defense of being over-analytic (psycho-therapists) to avoid the False Core Driver.

2) Pathologizing: An action of the False Self of projecting a psychological pathology onto another, so that it feels less inadequate, worthless (False Core), etc.

Chapter IX

Hypervigilance, the psychological component of the biological searching mechanism, is rooted in the nervous system's fight/flight overgeneralizing mechanism.

This process, aberrated with past associations, is not in present time since it was organized by the child to handle the outer chaos of mom/dad/society, etc.

This age-regressed tendency is an attempt to organize (mis)perceived chaos in present time. This mechanism runs on automatic.

Hence, over-analyzing and typing mechanisms re-enforce this age-regressed tendency.

EXTREMELY IMPORTANT

As with all False Core Drivers, reasons are created as to why I feel (*fill in the blank*) to justify the "feelings" and to give a way-out (i.e., a path or solution to this false reason). In this way it must be understood that you have come up with a *false* reason to explain why the separation occurred. The *False* conclusion is the False Core and it is solidified in theory at age 5-12 months. It is a false and inaccurate assessment of the natural process of separation. The first False conclusion assumes that if only I were not (*fill in the blank*), or if only I were (*fill in the blank*), the inevitable separation would not have occurred.

What you are doing is trying to solve the problem of separation by imagining a reason for it ("you"). But your conclusions are False, as to why you are separate. The False conclusion is an attempt to give a reason for the separation (which is inevitable). The *False Self Compensator must therefore be a False solution since it is based on a False conclusion.* In this way the False solution is erroneous and false because it is based on a false and erroneous conclusion drawn from the shock of the Realization of Separation, by an infant.

QUANTUM PSYCHOLOGY PRINCIPLE:

From the erroneous reasons and conclusions of the False Core come solutions which are erroneous and false because they are based on erroneous conclusions.

THE REASONS FOR I-DENTITY

I am inadequate because (*fill in the blank*), i.e., I am stupid, I was cursed, I am empty, etc.

FALSE CORE DISTRACTORS

The False Core Driver has a distraction of melancholy, depression, jealousy, envy, abandonment and betrayal. These distracting states are feelings associated with the erroneous reason and conclusion of the False Core which has a masochistic tendency. In other words, in order to a-void the pain of "I am inadequate," people can endure, put themselves through and create pain in themselves and others. The False Self Compensator can become over-analytic or overly reasonable to prove how adequate and reasonable it is. In short, *distractors are anything you do to distract yourself from the False Core Driver*.

As mentioned earlier, the mind moves very rapidly and people are always trying to get out of their False Core Driver. But the story or explanation you tell yourself and others as to why you feel what you feel is also a distractor. The story comes after the non-verbal experience and is an attempt to justify the "I am inadequate." But the False core has nothing to do with "my mother didn't love me" or "my father kicked me"; in other words it has nothing to do with the story. For example, I had a client who told me, "I am the way I am because my father worked seven days a week and my mother worked with him six or seven days a week. That was his story. During one session, I met his parents and this client said to his mother, "God, you worked when I was a little kid. You worked everyday. I never saw you." She turned to him and said, "I worked two days a week." *The story comes after the experience and it is a justification and a distraction* for the False Core and resisting the False Core, respectively

False Core Distractor Inner States:
Depression, Betrayel, Jealousy and Abandonment

Identities
Ichazo I-dentity #1—Smart
Ichazo I-dentity #2—Stupid

Ichazo calls smart the "over-reasoner." Quantum Psychology would say the "overly-reasonable."

The Quantum Psychology Identities:
False Core Driver: I Am Inadequate
False Self Compensator: Trying to be *overly* adequate

by being super reasonable by being over analytic and being able to figure things out.

AN ESSENTIAL COMPARISON

All the False Core Drivers experience some form of a lack. Why? Because unknowingly they are all comparing themselves to **ESSENCE**. In other words, "I'm imperfect," "I'm worthless," "I'm unable to do," "I'm inadequate" are all in comparison to **ESSENCE** (see Volume III).

It should be further noted that the issue of a lack can never be handled at a psychological level.[1] Why not? Because the I-dentity at first *appears* to be facing outward toward the world. It judges itself against others, comes up short and feels inadequate. In reality, the *I-dentities* are facing inward, comparing themselves to **ESSENCE.** *The truth is that I-dentities are less than or have a lack compared to* **ESSENCE**. For this reason, the way to handle the issue of a lack is to complete the I-dentity process by reabsorbing the False Core Driver and False Self Compensator *I-dentities* back into **ESSENCE**—and then experiencing its essential quality.

HEALING STYLE:
Merger through being "adequate," reasonable, figuring people out, over-analysis and over-diagnosis, over-typing.

PROTOCOLS FOR DISMANTLING THE FALSE CORE-FALSE SELF

These next questions should be considered as mini-meditations. The purpose of the second pair of *I-dentities* is to Ask the I-dentity a question and notice what, if anything, pops up.

---

[1]This is discussed more deeply in The Tao of Chaos, Quantum Consciousness, Volume II.

Chapter IX

FALSE SELF COMPENSATOR—
"I have to prove I am adequate" I-dentity

1. Where do you experience the *"I have to prove I am adequate" I-dentity* in your body? Write down the answer until nothing pops up.

2. Who modeled the *"I have to prove I am adequate" I-dentity?* Ask the I-dentity this question and write down the answer until nothing pops up.

Therapeutic Note

At this point, it is advantageous to externalize the *I*-dentity. This can be done in the following way:

> Step I: Notice where in the present time body the *I*-dentity is located.
> Step II: Notice the *I*-dentity's size and shape.
> Step III: Take the label off of the *I*-dentity. Have it as energy.
> Step IV: Allow the *I*-dentity as energy to move from the present time body to another physical location in the room, and then to become solid again.

3. Has the *"I have to prove I am adequate" I-dentity* ever been used as a way of resisting feeling betrayed? (This question is phrased this way because often the *"I have to prove I am adequate" I-dentity* is used to resist experiencing another emotion, for example, betrayal, abandonment, jealousy, depression or melancholy.) Ask the *I*-dentity this question and write down the answer until nothing pops up.

4. Has the *"I have to prove I am adequate" I-dentity* ever been used as a way of trying to overcome feeling abandoned? Ask the I-dentity this question and write down the answer until nothing pops up.

5. Has the *"I have to prove I am adequate" I-dentity* ever been used as a way of trying to overcome feeling depressed? Ask the *I*-dentity this question and write down the answer until nothing pops up.

6. Has the *"I have to prove I am adequate" I-dentity* ever been used as a way of trying to overcome melancholy? Ask the *I*-dentity this question and write down the answer until nothing pops up.

7. Has the *"I have to prove I am adequate" I-dentity* ever been used as a way of trying to overcome feeling inadequate? Ask the *I*-dentity this question and write down the answer until nothing pops up.

8. Has the *"I have to prove I am adequate" I-dentity* ever been used as a mask or False Self so you did feel inadequate? Ask the I-dentity this question and write down the answer until nothing pops up.

9. Has the *"I have to prove I am adequate" I-dentity* ever got into betrayal, abadonment, jealousy, depression or meloncholy memories and used the *"I have to prove I am adequate" I-dentity* as the only way of having a relationship with another. (Look for Mom and Dad). Ask the *I*-dentity this question and write down the answer until nothing pops up.

10. During the emergence→*I*-dentification of this *"I have to prove I am adequate" I-dentity*, which hides inadequacy, was there any chaos that occurred? Ask the I-dentity this question and write down the answer until nothing pops up.

11. Ask the *"I have to prove I am adequate" I-dentity* to recall a time when it had to be separate from mother, and to overcome this feeling it became *super* adequate. Ask the I-dentity this question and write down the answer until nothing pops up.

Chapter IX

12. Notice how *"I have to prove I am adequate" I-dentity* resists and defends against the shock of the Realization of Separation.

13. Has the *"I have to prove I am adequate" I-dentity* ever been used as a way of getting another to relate to it? Ask the *I*-dentity this question and write down the answer until nothing pops up.

14. During the process of taking on and using this *"I have to prove I am adequate" I-dentity*, were there any assumptions, beliefs, or decisions that the *I*-dentity made? Ask the *I*-dentity this question and write down the answer until nothing pops up.

15. During the process of taking on and using this *"I have to prove I am adequate" I-dentity* were there any fantasies the *I*-dentity had—e.g., How can I get or have *(fill in the blank)*. "Someday someone will come and take care of me or see how adequate I am." Ask the *I*-dentity this question and write down the answer until nothing pops up.

16. Has the *"I have to prove I am adequate" I-dentity* talked to itself (i.e., told yourself stories) to increase the suggestion of adequacy? Ask the *I*-dentity this question and write down the answer until nothing pops up.

17. Notice, again, where in your body you experience the *"I have to prove I am adequate" I-dentity*. Write down the answer until nothing pops up.

18. Peel back the *"I have to prove I am adequate" I-dentity* and notice how it might be covering a feeling of emptiness (as in a lack) in your body.

19. Keep peeling off any labels that might cover the **SPACIOUSNESS** and look out at the emptiness label and the *"I have to prove I am adequate" I-dentity*.

20. Ask the empty label and the *"I have to prove I am adequate" I-dentity*, "What are you seeking more than anything else in the world?" Ask them this question, allow them to answer and write down the answer until nothing pops up.

21. Experience the essential quality that the empty label and the *"I have to prove I am adequate" I-dentity* and any coverings (labels on **SPACIOUSNESS**) are seeking.

22. Notice the size and shape of the empty label and the *"I have to prove I am adequate" I-dentity*. Notice the **SPACIOUSNESS** they are floating in. See the layers and the **SPACIOUSNESS** as the same substance, while you continue to feel the essential qualities of **ESSENCE**.

FALSE CORE DRIVER—
I Am Inadequate I-dentity

Notice where in your body you experience the *"I am inadequate" I-dentity*. Go into the posture of the "I am inadequate" I-dentity and notice what happens to your body. Go into the "I am inadequate" I-dentity posture and see if it is a physically depressed I-dentity.

Therapeutic Note

At this point it is advantageous to externalize the I-dentity. This can be done in the following way:

Step I: Notice where in the present time body the *I*-dentity is located.
Step II: Notice the *I*-dentity's size and shape.
Step III: Take the label off of the *I*-dentity. Have it as energy.
Step IV: Allow the *I*-dentity as energy to move from the present time body to another physical

location in the room, and then to become solid again.

1. Has the *"I am inadequate" I-dentity* ever been resisted by trying to *figure out* what was going on emotionally for you or another? Ask the *I*-dentity this question and write down the answer until nothing pops up.

2. Has the *"I am inadequate" I-dentity* ever been used to justify getting therapy or seeking a spiritual practice? Ask the *I*-dentity this question and write down the answer until nothing pops up.

3. Whether you are a therapist or not, have you ever trance-ferred your *"I am inadequate" I-dentity* onto a client (or another) and then felt really adequate helping the other to analyze their way out a problem? Ask the *I*-dentity this question and write down the answer until nothing pops up.

4. As a child, did you ever have a parent who felt inadequate and depressed who you had to over-analyze, figure out or become super-adequate for (i.e., be their parent)? Ask the I-dentity this question and write down the answer until nothing pops up.

5. Has the *"I am inadequate" I-dentity* ever been used to justify or explain why someone might want to be separate or not respond to you? Ask the *I*-dentity this question and write down the answer until nothing pops up.

6. Prior to your taking on the *"I am inadequate" I-dentity*, was there any chaos or shock that occurred? Write down the answer until nothing pops up.

7. Has the *"I am inadequate" I-dentity* ever been used as a way to make you feel special, different or more spiritual from another? Ask the *I*-dentity this question and write down the answer until nothing pops up.

8. Has the *"I am inadequate"* I-dentity ever been used to help a parent (i.e., they felt that way so you matched them to have a relationship)? Ask the *I*-dentity this question and write down the answer until nothing pops up.

9. Has the *"I am inadequate"* I-dentity ever blamed itself for being not smart enough? Ask the *I*-dentity this question and write down the answer until nothing pops up.

10. Has the *"I am inadequate"* I-dentity ever been used to make itself feel wrong and "beat itself up" or criticize itself? If yes, whose I-dentity (mom or dad got internalized). Write down the answer until nothing pops up..

11. During the process of "taking on" this *"I am inadequate" I-dentity*, what decisions, beliefs or assumptions were made? Ask the *I*-dentity this question and write down the answer until nothing pops up.

12. During the process of taking on the *"I am inadequate" I-dentity*, what fantasies were created? Ask the *I*-dentity this question and write down the answer until nothing pops up.

13. Ask the *"I am inadequate"* I-dentity to recall a time in your early childhood where it felt abandoned because mom or dad left. Ask the *I*-dentity this question and write down the answer until nothing pops up.

14. Ask the *"I am inadequate"* I-dentity to recall a time in present time where it felt abandoned because a lover or friend was unavailable. Ask the *I*-dentity this question and write down the answer until nothing pops up.

15. Ask the *"I am inadequate"* I-dentity to recall a time where it felt betrayed by mom or dad. Ask the *I*-dentity this question and write down the answer until nothing pops up.

Chapter IX

16. Notice if the *"I am inadequate"* age-regressed I-dentity "mind reads," hallucinates or imagines what another's problem is, so that it could fix it with their over-analytic abilities? Ask the *I*-dentity this question and write down the answer until nothing pops up.

17. Has the *over-analytic I-dentity* ever been used to care take another person? Ask the *I*-dentity this question and write down the answer until nothing pops up.

18. Notice where, again, in your body you experience the *"I am inadequate"* I-dentity.

19. Peel back the depression or empty label, etc., and the *"I am inadequate"* I-dentity and notice that is underneath it.

20. Notice the **SPACIOUSNESS** that is underneath all of that.

21. Be in the **SPACIOUSNESS** and notice the labels the I-dentities, etc., are floating in.

22. Ask the I-dentities from the **SPACIOUSNESS**, and allow them to answer, "What are you seeking more than anything else in the world?" Ask the I-dentity this question and write down the answer until nothing pops up.

23. Experience that essential qualities and the **SPACIOUSNESS** that those I-dentities and labels are seeking.

24. See the size and shape of those labels and the two *I-dentities* and the **SPACIOUSNESS** that surrounds them as being made of the same substance, while you continue to feel the essential quality inside your physical body.

Demonstration

Carla is a 53 year old psychologist from England. We started with the distractors for "I am inadequate," i.e., *depression*.

Wolinsky: As a therapist, have you ever transferred your depressed I-dentity onto a client and then tried to help them to over analyze *their* way out of it?

Carla: Yeah, I think there is a tendency to do that sometimes.

Therapeutic Note

Here we see a classic counter-tranceference. The therapist transfers the depressed I-dentity on a client and then becomes overly reasonable or overly analytic.

Wolinsky: What are you feeling right now?

Carla: I feel nervous, anxious.

Wolinsky: What's the worse of this anxiety?

Therapeutic Note

By staying with "what's the worst of it," the False Core driver can be revealed. (See "Getting Your False Core.")

Carla: I freeze, blank, go numb.

Wolinsky: So by going numb, what experience are you resisting?

Carla: My inadequacies.

Wolinsky: So is there a layer called "I'm inadequate," and there's a layer of depression to hide the inadequacy and then there's a layer of being over reasonable to handle the depression?

Carla:	Yes, I experience that.
Wolinsky:	As a child did you ever have a parent who felt depressed or inadequate? And you had to be overly reasonable with them to try to get them out of it?
Carla:	My mother was depressed. Looking back at her, I see that she was depressed a lot of the time. My role was in being reasonable.
Wolinsky:	How about using reason to try to help you reason your way so you could survive in that context?
Carla:	Yes, that's more like it.
Wolinsky:	What I would like for you to do is have a depressed mother (force) over there and to create in response a very reasonable child (counter-force) which had to be reasonable to survive.

Therapeutic Note

Here we are using force theory and separating *the reasonable* child I-dentity from her as well as the depressed mother. In other words, the depressed mother is also not her.

Wolinsky:	Where in your body do you experience the kind of depressed, inadequate little girl?
Carla:	In my trunk through here. In my chest, in my abdomen.
Wolinsky:	In your core. Tell me a difference between you and this depressed little girl who feels inadequate.
Carla:	I can be spontaneous. And that feeling doesn't end.

Wolinsky: Have you ever used this over reasonable I-dentity to figure out why somebody might be separate or not respond to you?

Carla: Well, in my family situation, I compared myself to my brother and sister who were much older and smarter than I was and very gifted and bright people.

Therapeutic Note

Notice how we can easily get into the Ichazo distractor "I am dumb, they are smart."

Wolinsky: Prior to you taking on this overly reasonable I-dentity to try to figure out why you felt inadequate (the False Core driver), was there any chaos or shock that occurred?

Carla: Yeah, first I was born and I think my mother had doubts about wanting to have a child.

Wolinsky: What are you feeling right now?

Carla: I feel sad and I feel like sinking.

Wolinsky: Be sad and sink.

Carla: I don't want to do that.

Wolinsky: What happens?

Carla: It's too much.

Wolinsky: What would you rather do?

Carla: I would rather be reasonable. I feel that being reasonable pulls me out. It's almost like a balance but it is going from sinking in my chest to my head. If it is a balance.

Therapeutic Note

Balance is what the "I cannot do" False Core is seeking.

Wolinsky: So you have a whole internal reasonable world that nobody knows anything about?

Carla: Absolutely.

Wolinsky: Create the sad and intensionally change it to reasonable to defend against the sad a few times.

Carla: That feels very familiar.

Therapeutic Note

We change one feeling into another to cover them up.

Wolinsky: How are you feeling now?

Carla: When you said that, I was very aware of my isolation.

Therapeutic Note

Here she moves to the isolation of the False Core driver "I don't exist" to defend against the sadness which covers the "I am inadequate." There we have two distractors.

Wolinsky: What's the worst of the isolation?

Therapeutic Note

There I work backward from distractor of isolation from the "I don't exist" False Core to the False Core driver of "I am inadequate."

Carla: Not knowing how to share it.

Wolinsky: What's the worst of that?

Carla: I don't know how to share.

Wolinsky: Now, look at me and say that again.

Carla: *I don't know* how to share.

Therapeutic Note

It is important to have the client make contact with you (another human being) when they make the statement.

Wolinsky: What's the worst of not knowing?

Carla: I am inadequate.

Wolinsky: Have you ever used the over reasonable (over adequate) to make you feel like a separate self or an individual?

Therapeutic Note

Here we see that to differentiate from mom, she had to dissociate (not feel) and overcompensate, be extra reasonable.

QUANTUM PSYCHOLOGY PRINCIPLE:
To differentiate, you often times overcompensate.

Carla: Yes, I have, in fact I have probably tried to do that a lot and tried in a sort of a way to be different.

Wolinsky: Have you ever used the over reasonable False Self[2] to avoid being overwhelmed by your own feelings?

---

[2] I must prove I am adequate by being overly reasonable.

Chapter IX

Carla: Yes, feeling sad, a deep sadness.

Wolinsky: Where do you experience the depressed inadequate I-dentity in your body?

Carla: In my chest, and of course (my chest), is like a band and it moves.

Wolinsky: Who modeled this depressed inadequate I-dentity for you?

Carla: My mother though she was not overt about it either.

Wolinsky: You mean you are not? (sarcastic)

Carla: I contain it well. I did something.

Wolinsky: You do contain it well.

Carla: Even for me.

Wolinsky: Tell me something you have hid from yourself?

Carla: That I am more capable than I think I am.

Wolinsky: This False Core that you are describing of depressed and inadequate, how does it seem to you now?

Carla: A little more distant.

Wolinsky: Have you ever used the depressed I-dentity as a way of resisting feeling abandoned? And tell me someone who abandoned you?

Therapeutic Note

Here we are going for the distractors—making them more explicit (implicit-explicit)

Carla: I believe my mother did.

Wolinsky: What did you create in response to that?

Therapeutic Note

Force theory, see Chapter XVII.

Carla: What I think is a lot of confusion and out of that I had to find a way to be and that was my world. I defined my whole being.

Wolinsky: Did you ever create confusion and feeling stupid (a distractor) to not feel your mother's depression?

Carla: Yes. That's what I do and then I try to figure my way out of it.

Wolinsky: Did you ever take on your mother's depression so that you could have a relationship with her and not have to feel the abandonment?

Therapeutic Note

Here we see a fusion where someone takes on Mom's I-dentity which includes her depression as the *only* way to have a relationship with her. In other words, "I will do anything, take on anything" to avoid the shock of the Realization of Separation.

Carla: Yes.

Wolinsky: Did you ever take on your mother's depressed in-adequate I-dentity so you would not have to feel her abandonment of you, so you would feel like she is always with you?

Chapter IX

Carla: Yes, I wanted to and I have tried to find other people to do that with.

Therapeutic Note

People hold on tenaciously to their psychological stuff with mom and dad so they do not have to feel the shock of the Realization of Separation.

Wolinsky: Did you hold your mother (internal image) with you at the same time you were trying to find other people?

Carla: Yes, I did.

Wolinsky: So hold onto your mother there and try to find other people so you don't have to deal with this. Do it now intentionally. What are you feeling now?

Carla: It feels scary.

Wolinsky: To be without your mother there?

Carla: No, she's there.

Wolinsky: She's there. Well, at least you are in relationship— you didn't like me saying that.

Therapeutic Note

We take on our parents' False Core to have a relationship.

Carla: Uh huh.

Wolinsky: What about it didn't you like?

Carla: I don't want to be with her.

Wolinsky: So it sounds like you are wanting to be with her and resisting being with her simultaneously.

Carla: Yes.

Wolinsky: If this reasonable I-dentity and if this depressed melancholy I-dentity (melancholoy is another distractor for "I am inadequate,"), if they were no longer there, what would you be afraid you would have to be responsible for?

Carla: Whatever is being experienced.

Wolinsky: Would you be willing to say I keep the melancholy mom inside of me and take refuge in her?

Therapeutic Note
Mom is placed inside and later spiritualized so as not to have to feel the shock of the Realization of Separation.

Carla: I take refuge in the melancholy mom inside of me. I still feel resistance.

Wolinsky: That's okay. How about this. I don't want to acknowledge that I do take refuge with the melancholy mom inside of me. Can you say that?

Carla: I don't want to acknowledge that I feel—I don't want to acknowledge that— Well, I can't acknowledge that I resist it.

Wolinsky: Take refuge in mom.

Carla: I don't want to acknowledge that I take refuge in my melancholy mother inside of me.

Chapter IX

Therapeutic Note

In Buddhism, they "take refuge" in the Buddha. Empowerment ceremonies in Buddhism, which people do again and again, is when you place a certain "Buddha" inside of you, worship "it," etc. You can easily see how it would be possible for people who are resisting the shock of the Realization of Separation to use this technique. Simply put, infants take on idealized images of mom and dad and keep them in their psyche to avoid the shock of the Realization of Separation.

There are many *spiritualized* approaches who work with inner images, the inner (*fill in the blank*). This *spiritualization* can be used as a defense. Furthermore, it helps to answer the question, why does psychotherapy or spirituality take so long or not work? If you are using spiritualized images from the past to defend against something (realization of separation), then psycho-spirituality can be used as a defense, i.e., if I give up my images and stuff with mom and dad, then I will have to deal with 1) the shock of the Realization of Separation, and 2) who am I now?

"Spiritual systems" who not only install an inner (*fill in the blank*), but suggest they (the images) are God should first resolve the transpersonal transference because otherwise there is a danger of placing yet another layer image, etc. on top of making mom into God. Also, "spiritual systems" who suggest you are and should be the "inner images" (*fill in the blank*) are supporting a fusion with mom (spiritualized mom). And finally, consulting with, praying to, and receiving answers from an "inner guide" can be a *spiritualized* tranceference created out of the infantile imagining and wishes and resistance to mom as she is (i.e., making mom into omniscient or a magical mom who gives me what I need, etc.).

Wolinsky: How are you feeling now?

Carla: Clever.

Wolinsky: Prior to you taking on the melancholy mom putting inside you so you could have a relationship,

	prior to you doing that, how big was your consciousness?
Carla:	It was big.
Wolinsky:	And after you took on this melancholy mom and put it inside so you could have a relationship so you didn't have to feel abandoned, how big was your consciousness?
Carla:	It shrank to a pea.
Wolinsky:	How are you feeling now?
Carla:	I feel quite clear.
Wolinsky:	Where in your body do you feel inadequate?
Carla:	It's a low feeling.
Wolinsky:	You feel it in a lower part of your body.
Carla:	Yes.
Wolinsky:	Take the inadequate experience and intentionally change it into depression to resist the inadequaste.
Carla:	Yes.
Wolinsky:	Does that feel familiar?
Carla:	Yes.

Therapeutic Note

Since depression is the distractor, "inadequacy" can be changed into depression automatically to a-void the "inadequacy."

Wolinsky:	Now, take the inadequate, change it into depression and decide there must be something wrong with me. Is that familiar?
Carla:	Very definable steps, Yes.
Wolinsky:	So do that again intentionally and slowly with awareness. Feel inadequate, change it into depression and then decide there is something wrong with me. How do you feel now?
Carla:	I never moved from it. It's good to have it clear. It's almost satisfying.
Wolinsky:	Do it one more time. Create the inadequate, change it into depression and then decide if there is something wrong with me.
Carla:	Uh huh.

Therapeutic Note

We had the False Core driver of *inadequate*, it shifted to depression (distractor), and then moving to "There's something wrong with me." (Another False Core.) If you do not understand this movement as a defense, then you could spend the rest of your life looking to find *what's wrong with me*, finding something and trying to *fix it*. In other words, you could spend all your time looking for *what's wrong with me* as a way of a-voiding the inadequate. It is important to see the *whole* pattern and the *False Core Driver* so you can develop more awareness around it and, hence, be free of it. Nisargadetta Maharaj says, "What you don't know, you are a slave of. What you know about, you can be free of."

Wolinsky:	Have you ever taken-on the depressed I-dentity of another person as a way of having a relationship with them or helping them?

Carla: Taking on somebody else's depression as a way of relating. Yes. I've done it with friends. I did it with my mother.

Wolinsky: Yes. How are you feeling now?

Carla: Profound relief.

Therapeutic Note
Taking on someone's depression to have relationship. The acknowledgment of *what is* yields relief. The denial of this yields grief.

Wolinsky: Have you ever taken on your mother's melancholy as a way of helping her?

Carla: Yes.

Therapeutic Note
Taking on someone else's pain is, from the child's mind, is a way of helping them. (From the point of view of structural and strategic family therapy.)

Wolinsky: Can you say "I took on my mother's depression and melancholy as a way of helping her and having a relationship with her?"

Carla: I took on my mother's melancholy and her depression as a way of helping her and a way of having a relationship with her.

Wolinsky: In your life did you ever have people around who were melancholy and depressed that you could help and reason it out?

Carla: Yes, I think it is a huge part of my life.

Chapter IX

Wolinsky: Recall a time in your early childhood when you felt abandoned because mom didn't connect with you?

Carla: I have lots of memories of my mother going away for periods of time and returning and I wanted her to really notice me at her leaving and when she came back. I remember really anticipating her coming back and that she would pay attention to me or notice me or be excited to see me and it didn't happen.

Therapeutic Note

Neglect is a force *not* coming at you.

Wolinsky: What did you create in response to that neglect? (Counter-Force).

Carla: I didn't exist, and then I just went into my own world.

Wolinsky: Which means you went internal?

Carla: Yeah,.

Wolinsky: And felt?

Carla: Well, at times, I didn't feel at all.

Wolinsky: Did you feel you didn't exist?

Carla: Yes, but I would *create* a physical world around me that made me feel that I *existed*.

Wolinsky: The way you resisted neglect and nonexistence was through creating an inner world to prove you existed.

Therapeutic Note

Moving from the "I am inadequate" False Core to "I don't exist" then "I exist," to distract from the False Core.

Carla: Yes.

Wolinsky: How are you feeling now?

Carla: I feel sad about that. Again it's that feeling of isolation and numbness and wanting to connect with her.

Therapeutic Note

Isolation is a by-product of the False Core "I don't exist," and wanting to connect is the False Self solution to the False Core of "I am alone."

Wolinsky: So feel a sense of being trapped in your own inner world. Does that feel familiar?

Carla: Absolutely.

Wolinsky: Okay, how are you doing now?

Carla: Relief again. I am experiencing **SPACIOUSNESS**.

Wolinsky: Inside of that **SPACIOUSNESS**, how does the inadequate I-dentity and the over reasonable I-dentity, how does it seem to you now?

Carla: They feel distant.

Wolinsky: Have your mother over there (in another location *in the room*), and see her inadequate False Core

	inside of her. How does this False Core driver that is now *inside your mother* seem to you now?
Carla:	It's there, I can see it but I don't feel pulled to it or like putting it into me or taking care of it. It's neutral.
Wolinsky:	Now, I am going to ask you this but I am going to ask everybody in the group this too, "Is it okay with you to allow your mother to have her pain?

Therapeutic Note

Generally it is *not* okay to allow your mother to have her pain. Therefore, to help her, we might take on her pain.

Carla:	Is it okay with me? I feel my mother's sadness.
Wolinsky:	Where in your body do you feel your mother's sadness.
Carla:	In my chest.
Wolinsky:	Take the label off of the sadness and allow that energy to go back to your mother. Now, from the space, I would like for you to look at your mother's False Core and to see into the **SPACIOUSNESS** just beyond her False Core into her **ESSENCE**.
Carla:	I see her False Core and the **SPACIOUSNESS** of her **ESSENCE**.
Wolinsky:	Can you have a relationship with "her" **ESSENCE** to "your" **ESSENCE**, beyond her False Core?

Therapeutic Note

Biological and psycho-emotional separation are inevitable. However, at the level of **ESSENCE**, there is a relationship beyond personality. This is differentiation within unity (**ESSENCE** to **ESSENCE** relationship).

Carla: I can right now.

Wolinsky: What is that experience like right now?

Carla: It's a wonderful feeling.

Wolinsky: Now, knowing that there is this relationship, "her" **ESSENCE** to "your" **ESSENCE**, would you be willing to allow *her* to have her pain and *her* False Core?

Carla: Absolutely.

Wolinsky: How are you feeling right now?

Carla: Open, connected, separate, free.

Therapeutic Note

It is a two step process. 1) letting mom have her False Core; 2) keeping the **ESSENCE** to **ESSENCE** relationship. Children believe the best way to help their parents is to take on their False Core. You can have the **ESSENCE** to **ESSENCE** relationship, but not "take on" the False Core of your mother in order to have a relationship. The relationship you really want is **ESSENCE** to **ESSENCE**. However, when you can't have that because of the Shock of the Realization of Separation, you will take on her False Core-False Self and have a False Core to False Core relationship because it is the only option, and merging is a biological function. Later, this False Core-False Self relationship gets "acted out" again and again in future relationships.

Chapter IX

DEMONSTRATION WITH RESISTED STATES:

Ellis: I think I was on four.

Wolinsky: Do you want to do a few minutes with me?

Ellis: Sure.

Wolinsky: Notice where in your body you first tend to feel inadequate.

Ellis: In the back of the head.

Wolinsky: So intentionally create that.

Ellis: Okay.

Wolinsky: Good, create it again.

Ellis: Okay.

Wolinsky: What happened?

Ellis: It just popped because I realized I was very adequately creating inadequacy. I was being a little too adequate in my inadequacy.

Wolinsky: (To group) So another state that a four would resist would be depression. (To him) So where in your body do you feel depression?

Ellis: More in my stomach.

Wolinsky: Good, create it again. Create it again. How does that seem to you now?

Ellis: Good.

Wolinsky: (To group) Another resisted state would be the experience of being stupid or dumb. (To him) Where in your body do you get the stupid/dumb experience?

Ellis: In my left side.

Wolinsky: Create it. Since you know your body, notice and track it through your whole other side, which would be your left side. Notice the impact of that and create it.

Ellis: Okay.

Wolinsky: Create it again. (Pause) Create it again. (Pause) Create it again. (Pause) How does that seem to you?

Ellis: Pretty easy.

Wolinsky: How are you doing?

Ellis: Much clearer.

Therapeutic Note

Step I: Go with a partner over all of the False Core Drivers in each fixation.

Step II: "Where do you feel it in your body?" "I feel it in my chest, in my jaw, etc."

Step III: Notice the resisted experiences. So, for example, if you are a worthless, False

Chapter IX

	Core, create worthless, its distractor, dependence, etc. Create every state you absolutely loathe.
Step IV:	Have them create it knowingly, consciously, intentionally *without the intention of getting rid of it*.
Step V:	Ask them to tell you an idea they have about the False Core or its distractors.
Step VI:	Each time they answer, say, "That's an interesting idea."
Wolinsky:	(To the group) Part I is to go through with a partner all the False Core Drivers and distractors in each fixation. Then Part II notice the resisted experiences. Create every experience you absolutely loathe.

(Demonstration continued)

Ellis:	Inadequacy is something you should be able to correct, improve or fix.
Wolinsky:	That's an interesting idea. What happened?
Ellis:	It just went to an idea. It floated off.
Wolinsky:	Tell me another idea you have about inadequate.
Ellis:	They move so fast. First, I had an idea that it was a hopeless idea to even investigate.
Wolinsky:	That's an interesting idea.
Ellis:	I have the idea that it is a state that I want to avoid.

Wolinsky: That's an interesting idea.

Ellis: Just flashing on all the work, what an effort it is to avoid it.

Wolinsky: You have a lot of effort to avoid it?

Ellis: It's a lot of work.

Wolinsky: That's an interesting idea. Do you have a False Core?

Ellis: A name for it? I don't know.

Wolinsky: Do you have that idea that you have a False Core?

Ellis: I haven't thought about it using those words.

Wolinsky: What words would you use?

Ellis: The underlying state.

Wolinsky: That's an interesting idea. How are you doing?

Ellis: It's a body sensation. Then, when you say that's an interesting idea, it just goes out, away. It just goes body state, idea, head, out.

Wolinsky: So tell me an idea you have about depression?

Ellis: That it is a depressing thought.

Wolinsky: That's an interesting idea.

Ellis: That it is a genetic family thing.

Wolinsky: That's an interesting idea.

Chapter IX

Ellis: That one wasn't so interesting to me. (Laughter)

Wolinsky: Do you have an idea that you have generations?

Ellis: Yeah. Okay, it's an interesting generational idea.

Wolinsky: Tell me an idea you have about depression, anything else that might be interesting?

Ellis: Seems a little boring right now. I search for a little medication, herbs.

Wolinsky: Boring is kind of an interesting idea. How are you doing now?

Ellis: Yeah, boring is an interesting idea.

Wolinsky: And this idea you have about this stupid thing, tell me an idea you have about stupid.

Ellis: Just the words, I should have known better, I knew better. I should have known better. That one sounds pretty silly.

Wolinsky: So you can have an interesting idea that comes out from your mother, I am assuming.

Ellis: That's an interesting idea.

Wolinsky: How are you doing now?

Ellis: It stayed here rather than going to my mom.

Wolinsky: So do you want to keep that interesting idea?

Ellis: That one is slower moving. That's here.

Wolinsky: Well, intentionally create it here and decide that you like that one and create it again.

Ellis: Okay.

Wolinsky: Create it again.

Ellis: Okay.

Wolinsky: Now, stop creating it. (Pause) Now, create it. (Pause) Now, stop creating it. (Pause) How are you doing?

Ellis: Good—really clear and open.

FOR EVERYONE—HOMEWORK (WITH A PARTNER)

After you process through the major False Core Driver-False Self Compensators, take a look at the inner states (or distractors) that are resisted or have been used to resist the False Core. In the above example, resisted states were inadequate, depression and stupid. Create the resisted states several times and experience them as just "interesting ideas." Maintain an awareness of the *Non-verbal I AM prior* to the False Core. Always take off the label off of the False Core. Do not believe the False Core Driver. For example, if a client comes into your office and says, "I feel sad because my mother never loved me and I have been to fourteen therapists and they all agree that that's *why* I feel sad," then I have a real hard time to break that habit, because the story has been reinforced. Mom is not in the room. The story comes after the experience and serves to justify and re-enforce the False Core Driver. In other words, the story is just an interesting idea and that's all it is. Psychology is an interesting idea but it can entrap you. In this way, do not try to use the False Self to change the False Core. It is seductive. If you are not your False Core, then why would you want to change, heal, transform, convert, etc. You only try to do these things if you subtly believe the False Core is you.

Chapter IX

CONCLUSION

1. Without using your thoughts, memory, emotions, associations or perceptions, are you adequate, inadequate or neither?
2. Without using your thoughts, memory, emotions, associations or perceptions, what does adequate or inadequate mean?
3. Without using your thoughts, memory, emotions, associations or perceptions, what does empty mean?

**NOTICE THE NON-STATE
STATE OF I AM PRIOR
TO THE FALSE CORE**

CHAPTER X
THE WAY OF THE NON-EXISTENT

FALSE CORE DRIVER:
I Don't Exist, I am Nothing, I have Nothing.

FALSE SELF COMPENSATOR:
I must prove that I am something, have something and that I exist

The next False Core driver "I do not exist" is one of the most interesting and—sometimes in Quantum Psychology work shops—the most frequent. "I" have found this False Core-False Self to appear frequently in ashrams, meditation groups, therapy groups, and spiritual groups. According to Quantum Psychology there are two reasons for this frequency:

First, like all other false premises, this particular False Core Driver seeks to re-enforce itself. In other words, in many "Eastern" spiritual groups the philosophy of "there is no-self," "observe and watch," etc., re-enforces the most primal defense of this False Core Driver: *OVER-OBSERVATION*.

Second, if we explore Reich's and Lowen's theory, we find that this False Core Driver which can be likened to the schzoid in "character analysis," is developed in-utero. This means that this False Core occurs not only in an energetic-genetic level, but it can also be visually seen within the bony structure of an individual. From a Quantum Psychology perspective, this means that this False Core begins to solidify not at 5-12 months but, as Reich and Lowen suggest, in utero.

Because of its "early development," Quantum Psychology sees the obsessive-compulsive False Core Driver and False Self Compensator as being even more deeply embedded in the body than any of the False Core-False Selfs. The underlying, non-verbal, pre-representational structure of "I do not exist," along with the early pain which resides far outside of awareness, could explain its predominance in meditation groups which search for answers (knowing (about)) in Eastern philosphy and religion. Please note that this no-self of the East is true at the level of the **VOID OF UNDIFFERENTIATED CONSCIOUSNESS**, but untrue at the level of biopsychology, (to be discussed in Volume, III).

For that reason, it is easy for this False Core Driver of "I do not exist" to collapse levels and choose a Buddhist (no-self) or Advaita (non-duality), a no "I" approach which re-enforces the False Core Driver of "I do no exist."

Quantum Psychology's view on this is that any teacher who allows students to "stay around" and do a "practice" which re-enforces their False Core is like offering an alcoholic another drink. This also suggests that the "thinking" therapies of Ellis, Korsybski, psycho-analysis and many other therapies which support "thinking" and "observation," can oftentimes serve to re-enforce the over-thinking distraction if someone has this False Core of "I don no exist."

Quantum Psychology asks therapists and spiritual teachers to discern *who is in front of them*, not re-enforce the False Core, and when necessary, to *refer out* regardless of their economic loss.

In "my" understanding, teachers in the Sufi tradition always sent students where they needed to go. But now "teachers" often don't see the shortcomings of their own system which they are following. Possibly because the "I don't exist" teacher has his or her own schizoid structure, insight barrier or intellectual armoring which prevents "seeing" who's in front of them. In this way, they keep students who are not appropriate to "their" approach through Observer-False Core Driver-False Self Compensator re-enforcement not comprehending their own systems' and their own personal limitations.

Chapter X

The Reason Given:
I do not exist because (*fill in the blank*)—"There is no purpose," "I am nothing," "I am empty," "I don't know").

The False Core Driver of "I don't exist" "feels" like "I am nothing, I have nothing." This is the false sense of nothing (as in a lack), not the **BIG NOTHING**. Their behavior reinforces this or attempts to get out of it in some way. Sometimes False Core Drivers of "I do not exist" move to "inadequate" or "alone" in order to *feel something, be something, or have something*. This False Core Driver is in a dissociated state, hence, it is not a feeler but rather it *thinks feelings*. This could be because, as Reich and Lowen suggest, the schizoid develops in utero because of the mother's rejection. Hence, they dissociate early *becoming over-observers as a defense*. Furthermore, they can confuse the dissociative void for the **VOID** or Samadhi. This *blankness*, termed by this False Core-False Self as the **VOID**, is rather a blankness placed over a trauma and acts as a defense against feeling. Their defensive resistance to feelings is illustrated in the example below but please bear in mind, this is an extreme case.

I once knew a doctor who was anorexic-bulemic. He *spiritualized* his eating disorder and defined it as "purification." Wearing only white, this guru wannabe put many of his patients on fasts and purgatives. Whenever he felt anything, he would stop eating. He even went on a 40-day fast. When I asked him why, he said "Jesus did it, Moses did it." He said he wanted to become a breathatarian (someone who never eats but lives on air). I suggested he go into some form of body therapy like breathing work to get in touch with his feelings.

This reminds me of a story I heard the comedian Swami Beyondananda tell. He said, when he was in the Himalayas, he met many very spiritual people who were breathatarians. One day he walked into a restaurant for breathatarians. There was no food but the *atmosphere* was great. He noticed that everyone had a clothespin on their nose and he asked why? "We're fasting," someone said.

DISTRACTOR I-DENTITIES

The distractor i-dentities of the False Core Driver "I don't exist" are "I don't know" and "I have to know." To clarify, an "I am inadequate" might have an "I am stupid *I*-dentity" and try to over-analyze their way out of it. The False Core Driver of "I don't exist," however, *accumulates* information because they imagine they are nothing and have nothing. If I have "something" (i.e., information and ideas), "I exist." For this reason there is an obsessive-compulsive addiction to *knowing* and accumulating information to avoid the pain of non-existence.

> **SPIRITUAL DEFENSE:**
> Over-observering, I am nothing, there is no self, there is no-I.

THE INVISIBLE

This False Core can make itself quite invisible. Invisibility is a two-edged sword, however. They make themselves invisible to others as a defense but they become invisible to themselves and their own needs. This too is oftentimes *spiritualized* since many Eastern traditions suggests no needs and no desires are "spiritual."

MORE DEFENSES

The defense of the False Core "I do not exist" is the False Observer. It is a False observer because it was created by the nervous system as a defense against feeling. They re-enforce this and they tend to be drawn into observer meditations which support no feelings, or labeling no feelings as good. In insight (Vipassana) meditation oftentimes over-observers are drawn into and attracted to it because it re-enforces their defenses. For an "I don't exist" False Core to become a Buddhist—which emphasizes "I am nothing" or "I don't exist" and stresses observation—is disastrous because it re-enforces the over-observation defense of this False Core-False Self. In short, certain spiritual or psychological systems are good for certain False Core Drivers while others only re-enforce certain False Cores. For

Chapter X

this False Core Driver insight meditation re-enforces the distractor of the "False-observer" as cognitive therapy or psychoanalysis re-enforces the distraction of "over-thinking."

REJECTION

The "I don't exist" False Core driver also contains the schizoid structure of rejection. People with this False Core assume that they will be rejected, and instead they reject first. In short, they first reject the world (Mom) in order to avoid the world's (Mom's) rejection. They do this by either rejecting themselves, "part of themselves," or "their body." This tendency to dissociate and dis-own "their" world and body is a fusion with a rejecting mom and forces them more deeply into False-observation of the dissociated void or spiritual systems which deny or reject feelings, the body or the world.
This False Core Driver can trance-fer the perceived engulfing smother mother onto the world and then dissociate from the world and their partner or spouse. They often set themselves up for rejection and begin to *isolate*. They assume "you are smothering (or engulfing) me." In short, they *assume* engulfment and to defend they dissociate/isolate from others or they defensively observe and call it spiritual.

Consider the two following stories: I met an "I don't exist" person once who always assumed he was objectively observing me rejecting him. He once gave me a book about different spiritual teachers and I asked him what he thought of the author. "He seemed like he's looking for a Guru to reject," he said. He thought he was in observation but he was in the "I don't exist" False Core Driver's schzoid structure which sees rejection everywhere and sets itself up for it. I once had a client who would not keep her agreements with me or with her professional community. I told her she was re-enacting her False Core Driver's defense by setting herself up for rejection and isolation. But she didn't appreciate that her chronic set-up for rejection coupled with her isolationism was this False Core's obsessive-compulsive tendency.

The "I am nothing" of this False Core Driver is not the experience of the **BIG NOTHING**, it is the deficient "I'm nothing" of this particular False Core. The **BIG NOTHING** is experienced as neither "I am nothing" nor "I am not nothing," that I neither exist nor do I not exist without any issue about it. In other words, without using your thoughts, memory, emotions, associations or perceptions, ask yourself, "Do I exist, not exist or neither?" Without using your thoughts, memory, emotions, associations or perceptions, ask yourself, "What is existence?" But the small "I am nothing" carries all the charge with it—"I'm nothing, I'm invisible, I don't exist," etc. This is how the **BIG NOTHING** gets confused with the *small nothing*, especially in so-called spiritual meditation or Eastern religious groups.

To illustrate, you can see a lot of this False Core in Ashrams in India. The spiritual system reinforces and was probably chosen unknowingly by this False Core-False Self to re-enforce its psychological system. They appear detached but it is the detachment of the defensive observer in reaction to and resistance against "I don't exist." Helen Palmer, calls this personality type the Unenlightened Buddha. Quantum Psychology sees this developing because the "I have nothing and I am nothing" has to act "as if" because to themselves they are nothing and do not exist. Acting enlightened becomes a spiritual defense, hoarding everything (information, thoughts, students and even spiritual teachings) to distract themselves.

This False Core Driver can become an over-knower accumulating "knowledge" (information) and thoughts. This accumulated information is not analyzed like the "I am inadequate" False Core but is just collected as data. This False Core confuses thinking with experiencing. In 1986, I trained a woman psychiatrist from Utah who had a False Core Driver of "I don't exist." She had written a book describing her spiritual experiences which were identical to Swami Muktananda's in his work, *Play of Consciousness*. It turns out she was a disciple of Muktanandas and had read the book and *thought* she had had the (his) experiences. She was acting *"as if"* she were enlightened, hence, the unenlightened Buddha.

> A workshop participate once said, The tendency to accumulate knowledge goes on excessively for me but deep down, I feel like I don't know. Underneath I couldn't tell anybody what I know by what I have

accumulated because the *I have to know* is really *I don't know*. If somebody asks, "What have you read, what are you learning?" "I don't know" is the automatic response.

This False Core isolates itself, however, there is still a comfort in the deficient emptiness because it defends against feeling; hence, it feels familiarity. The disconnection proves they do not have anything and justifies isolating themselves from themselves and others, and then psychologically disappearing. As with all False Core Drivers, this False Core-False Self recreates and proves itself thus following Freud's repetition compulsion.

Quantum Psychology sees the False Core as a conclusion, an abstraction taken from a trauma, its repetition only re-enforcing itself as a self-preserving defense.

QUANTUM PSYCHOLOGY PRINCIPLE:
All decisions or conclusions are abstractions which are attempts to understand the chaos of a trauma. These conclusions act as a defense and only re-enforce themselves.

OVER-THINKING

"I have nothing"→"over-thinker" to fill in the space. The spiritual world of meditation is full of this False Core Driver because they can isolate and can go on a silent retreat, isolate and not have to talk to anybody which is *spiritualized* as spiritual. They are attracted to spiritual systems which encourage them to observe and isolate and, hence, to use meditation systems which justify (defensive) observation, silent retreats, isolation (withdrawal), nonexistence and dissociating from feelings, which is a *spiritualized* re-enforcement of this False Core-False Self.

If your False Core is "I don't exist," you might try to overcome it by being social (Ichazo) in the classic way or to resist, thus, seeming invisible. Or at some level, you might know that you are being social but deep down feel you don't exist. You can act "as if" you were social, while always being aware you *don't exist*. I had a friend with this False Core Driver who used to come visit me. He was the best house guest I ever had. In the house, he acted "as if" he were nonexistent. But if we went out to dinner, he would be flirting with the waitress and talking to the lady behind the cash register.

OBSERVATION IS NOT ESSENCE (SEE VOLUME III)

In **ESSENCE**, there is neither an observer-observed nor judgment evaluation or significance. In **ESSENCE**, there is observation with no object, there is no duality. This False Core Driver uses observation as a defense. The observer is part of the ego which is the glue holding the ego together and helping it mediate between the external (mom-dad, spiritual rules, society, etc.) and the Id, the repository of urges, sensations and emotions. Observation for this False Core demonstrates the layers of its defense. And so, observation is both a defense against "I don't exist" and—simultaneously, since there are no feelings about it—a re-enforcement of "I don't exist."

Simply put, in observation you don't have to *know or experience* the False Core of "I don't exist." Just as the "I don't know" is a defense against knowing and experiencing "I don't exist." This is why there is an *insight barrier*, the observations are *false* and you are not free to experience (or not experience) what is occurring. Instead, you are full of judgments (feelings are bad), evaluations (this makes me special or different from others) and a sense of significance (I am more spiritually evolved than others) (see Quantum Consciousness). But over-observing is a form of false detachment and a defense against feeling. Since this False Core is less aware of their body, it "chooses" spiritual systems which re-enforce its defense with slogans like, "Be detached" and "You are not your body."[1]

Healing Style:
Healing by merging through knowing, accumulating, isolating, remaining silent, withdrawing, thinking and acting "as if."

---

[1] I am not the body is true at the **VOID OF UNDIFFERENTIATED CONSCIOUSNESS** or **NAMELESS ABSOLUTE**. This "I Am Not" the body confuses the **VOID OF UNDIFFERENTIATED CONSCIOUSNESS** and **NAMELESS ABSOLUTE** with the Biological level and is gravitated to by this Observer-False Core-False Self complex.

The Antidote
As with all False Cores,
1. *Trace* it.
2. Notice it.
3. Experience it.
4. Without believing it, peel it back to your False Core ("I don't exist").
5. Experience it.
6. Be in the verbal **I AM**.
7. Then be in the non-verbal **I AM** prior to the False Core Driver, i.e., no thoughts, memory, emotions, associations or perceptions.

QUANTUM PSYCHOLOGY PRINCIPLE:
The False Core is a concept.

DISTRACTOR INNER STATES:
I feel nothing, over-thinking, they might engulf me, withdrawal into *spiritualized* silence.

THE PROTOCOLS FOR DISMANTLING THE FALSE CORE-FALSE SELF

QUANTUM PSYCHOLOGY'S FALSE CORE DRIVER:
"I do not exist."

FALSE SELF COMPENSATOR:
"I have to prove I exist."

DISTRACTOR I-DENTITIES:
I-dentity #1—"I do not know"
I-dentity #2—"I have to know"

I-dentity #1—I have nothing.
I-dentity #2—I must accumulate to be or have something, or I have to know.

ICHAZO'S I-DENTITY:
I-dentity #1, Social,
I-dentity #2, Antisocial

The following questions are mini-meditations. The purpose is to ask yourself a question and notice what, if anything, pops up.

FALSE CORE DRIVER—I DON'T EXIST
1. Notice where in relationship to your body you experience the *"I don't exist" I-dentity*. Ask yourself this question and write down the answer until nothing pops up.

Therapeutic Note

Notice that this question asks, where *in relationship* to your body rather than where *in* your body. The reason for this is that the "I don't exist" schizoid of this False Core Driver dissociates and loses awareness of the body. For this reason, they are over-observers. This is not "spiritual." As you might become aware of many "spiritual" seekers and heavy mediators have this False Core Driver. The automatic observation is a defense against feeling the shock of the Realization of Separation. The by-product is lost awareness of their body. For this reason, this False Core Driver-False Core Compensator adapts spiritual philosophies which reinforce their character structure of schizoid. They are in what Quantum Psychology calls premature observation. It is automatic observation without the ability to *be* a feeling or emotion. They *reject* it automatically and hence as Swami Beyondananda would say they experience premature rejection.

2. When you Identify with the *"I don't exist" I-dentity*, are you in your body, slightly separate from it, or around your body? Write down the answer until nothing pops up.

Chapter X

3.   Who modeled this *"I don't exist" I-dentity*? Write down the answer until nothing pops up.

Therapeutic Note

At this point it is advantageous to externalize the *I*-dentity. This can be done in the following way:

   Step I: Notice where in the present time body the *I*-dentity is located.
   Step II: Notice the *I*-dentity's size and shape.
   Step III: Take the label off of the *I*-dentity. Have it as energy.
   Step IV: Allow the *I*-dentity as energy to move from the present time body to another physical location in the room, and then to become solid again.

4.   Has the *"I don't exist" I-dentity* ever been used to prevent feelings (if so what feelings?) Look specifically for intimecy, closeness contact. Ask the *I*-dentity this question and write down the answer until nothing pops up.

5.   Has the *"I don't exist" I-dentity* ever been used so that people would not see that you have nothing or are nothing? Ask the *I*-dentity this question and write down the answer until nothing pops up.

6.   Has the *"I don't exist" I-dentity* ever been used as a way of rejecting itself? Ask the *I*-dentity this question and write down the answer until nothing pops up.

7.   Has the *"I don't exist" I-dentity* ever been used as a way of avoiding imagined rejection? Ask the *I*-dentity this question and write down the answer until nothing pops up.

8. Has the *"I don't exist" I-dentity* ever been used in relationship to another out of fear of them engulfing you? Ask the *I*-dentity this question and write down the answer until nothing pops up.

9. Has the *"I don't exist" I-dentity* ever become an observer as a way to a-void engulfment or rejection by mom, who it was imagined might have smothered you? Ask the *I*-dentity this question and write down the answer until nothing pops up.

10. Has the *"I don't exist" I-dentity* created an observer to watch as the only solution to having some semblance of a separate individual self? Ask the *I*-dentity this question and write down the answer until nothing pops up.

11. Has the *"I don't exist" I-dentity* and/or its dissociated silent observer ever been used to reinforce a spiritual path? Ask the *I*-dentity this question and write down the answer until nothing pops up.

12. Has this *I*-dentity ever thought it was special, different or knew something others did not know (i.e., that *you* don't exist)? Ask the *I*-dentity this question and write down the answer until nothing pops up.

13. Has the *"I don't exist" I-dentity* ever felt more evolved spiritually than another? Ask the *I*-dentity this question and write down the answer until nothing pops up.

14. Has the *"I don't exist" I-dentity* ever been used as a way to practice meditation or justify going on a silent retreat? Ask the *I*-dentity this question and write down the answer until nothing pops up.

15. Has the dissociated observer of the *"I don't exist" I-dentity* ever been used as a way of punishing another by withdrawing? Ask the *I*-dentity this question and write down the answer until nothing pops up.

Chapter X

16. During the I-dentification with the observer isolationist *I*-dentity, was there a chaos or shock that occurred? Ask the *I*-dentity this question and write down the answer until nothing pops up.

17. By taking on the *"I don't exist" I-dentity*, what was lost? Write down the answer until nothing pops up.

18. During the process of taking on the *"I don't exist" I-dentity*, what decisions, beliefs, or assumptions did it make? Ask the *I*-dentity this question and write down the answer until nothing pops up.

19. During the process of taking on the *"I don't exist" I-dentity*, what stories did it tell itself about how much better it was than other people because it had the ability to isolate, live alone, go into silence, or were "chosen" for this most important and evolved path? Ask *I*-dentity this question and write down the answer until nothing pops up.

20. Notice again where in your body you experience the *"I don't exist" I-dentity*.

21. Peel back that I-dentity and the empty label beneath it and notice the **SPACIOUSNESS**.

22. Go into the **SPACIOUSNESS** inside your body, and from inside there view the *"I don't exist" I-dentity* and the emptiness label and the dissociated observer.

23. Ask the *"I don't exist" I-dentity* and any labels or layers from back in the **SPACIOUSNESS**, what are you seeking more than anything else in the world and sllow the layers to answer. Ask the this question and write down the answer until nothing pops up.

24. Experience what the *I*-dentity, the layers, and the labels are seeking from inside the **SPACIOUSNESS** of **ESSENCE**.

25. Notice the size and shape of the i-dentities and labels, and the space that surrounds them, and see the i-dentities and the labels as made of the same substance as the **SPACIOUSNESS** while you experience that essential quality of **ESSENCE**.

POLARITY #2—I HAVE TO PROVE I EXIST

1. Notice where in your body you experience the *"I have to prove I exist"* I-dentity. Ask the I-dentity this question and write down the answer until nothing pops up

2. For a moment, take on the *"I don't exist"* I-dentity, and notice what's that like, and then very slowly move from the *"I don't exist" I-dentity* into the *"I have to prove I exist" I-dentity* and notice how one is in contrast to the other.

3. Who modeled this *"I have to prove I exist" I-dentity*? Ask the I-dentity this question and write down the answer until nothing pops up.

Therapeutic Note

At this point it is advantageous to externalize the *I*-dentity. This can be done in the following way:

Step I: Notice where in the present time body the *I*-dentity is located.
Step II: Notice the *I*-dentity's size and shape.
Step III: Take the label off of the *I*-dentity. Have it as energy.
Step IV: Allow the *I*-dentity as energy to move from the present time body to another physical location in the room, and then to become solid again.

4. Has this *"I have to prove I exist"* I-dentity ever been used as a way of getting something from somebody? Ask the *I*-dentity this question and write down the answer until nothing pops up.

Chapter X

5. Has the *"I have to prove I exist"* I-dentity ever been used as a way to hide your general irritation and rejection of people? Ask the *I*-dentity this question and write down the answer until nothing pops up.

6. Does the *"I have to prove I exist"* I-dentity ever act "as if" it was warm and engaging when actually you are disinterested and detached? Ask the *I*-dentity this question and write down the answer until nothing pops up.

7. Has the observer isolationist *I*-dentity ever been used and then hidden by pretending or acting "as if" you were present? Ask the *I*-dentity this question and write down the answer until nothing pops up.

8. Has the "I *have to prove I exist" I-dentity* ever been used as the only solution to the problem of hiding dissociation, lack of connection, and the feeling of "I don't exist"? Ask the *I*-dentity this question and write down the answer until nothing pops up.

9. By taking on the *"I have to prove I exist" I-dentity*, what general feelings towards other people are you resisting expressing or acknowledging? Ask the I-dentity this question and write down the answer until nothing pops up.

10. During the Identification with on this *"I have to prove I exist" I-dentity*, was there a chaos or shock that occurred? Ask the *I*-dentity this question and write down the answer until nothing pops up.

11. During the process of taking on this *"I have to exist" I-dentity*, what beliefs, decisions or assumptions did this *I*-dentity create? Ask the *I*-dentity this question and write down the answer until nothing pops up.

12. During the process of taking on this *"I have to prove I exist" I-dentity*, did this *I*-dentity ever check to see how other people

were acting, and act the way they act so that it could feel like it fit in. Ask the *I*-dentity this question and write down the answer until nothing pops up.

13. Has the (over-social) *I*-dentity ever been used to act "as if" it knew things it did not know? Ask the *I*-dentity and write down the answers.

14. Has the *"I have to prove I exist" I*-dentity ever been used to accumulated knowledge or information from other's behavior as " how to be, act, behave, feel or respond so that it could continue to build an "as if" personality and not have the *"I don't exist."* Ask the *I*-dentity.

15. Has the *"I have to prove I exist" I*-dentity, ever created theories to act "as if" it knew, when it did not? Ask the *I*-dentity and write down the answers.

16. Has the *"I exist" I*-dentity ever been used to imagine that another was going to reject it and so it projected *"I don't exist"* out on them? Thus proving it exists and they are nothing. Ask the *I*-dentity this question and write down the answer until nothing pops up.

17. Where in or in relationship to your body do you experience this integrated crystalized "as if," I have to act "as if" *I exist I-dentity* or I have to act "as if" "I know"To avoid "not knowing" and not exist? Ask the I-dentity this question and write down the answer until nothing pops up.

18. Peel back the *"I have to prove I exist" I*-dentity and notice the *"I don't exist" I*-dentity underneath it.

19. Notice if there is an empty label on top of the **SPACIOUSNESS** and peel the label back and go into the **SPACIOUSNESS** inside your body.

20. From inside the **SPACIOUSNESS** of **ESSENCE** inside your body, look out at the *"I have to prove I exist"* I-dentity, *I don't exist* and any labels.

21. Ask each one, " What are you seeking more than anything else in the world?" and allow them to answer Write down the answer until nothing pops up.

22. From inside the **SPACIOUSNESS** of your **ESSENCE**, experience the essential quality that the i-dentities and labels are seeking.

23. Notice the size and shape of the *I*-dentities and labels, see the *I*-dentities and labels as the same substance as the **SPACIOUSNESS**, and as you let them dissolve, continue to feel the essential quality that those *I*-dentities were seeking at the level of **ESSENCE**.

Demonstration
I don't exist (I am nothing)

Louise is a 46 year old therapist from Los Angeles.

Louise:	I'm afraid Stephen. I'm afraid you will see that I am nothing. I've got this image of a fire pole inside myself that I hang onto.
Wolinsky:	Where's the fire pole in your body?
Louise:	Here and in here. (Points to the middle of her body.)
Wolinsky:	Notice the size and shape of the fire pole.
Louise:	I'm shaking now. It feels more diffused than it used to. It's like a beam or rod in the center of me. It's disappearing while I try to find it.

Wolinsky:	Notice the size and shape of it. And realize that as you look at it, that it is an image you (your nervous system) created so that you would have or be *something* rather than *nothing*. Notice you are able to know it, and therefore, it cannot be you. It's not who you really are. How are you feeling?

Therapeutic Note

Please note: 1) Nisargadetta Maharaj's statement that, "anything you know about cannot be." You are beyond anything perceivable or conceivable And 2) from a therapist's point of view, Louise was a client I had known for years. I trusted her and she trusted me. So we were able to deconstruct the image of the pole she was hanging onto. With someone I didn't know as well, I wouldn't have done that.

Louise:	Shaky.
Wolinsky:	So I would like for you to make the statement, "I am nothing." "I have nothing."
Louise:	"I am nothing." There was feeling when you said "I have nothing."
Wolinsky:	What was the feeling?
Louise:	Like you just spoke the truth.
Wolinsky:	Can you say "I have nothing?"
Louise:	"I have nothing." This sounds awful to say. If I don't know you very well so I can sort of make you nothing. Then I sort of don't have to deal with all that shit going on.

Therapeutic Note

I have her make the statement (i.e., make the implicit explicit) so that she can own, accept and acknowledge her ideas about herself first before she can go beyond. She cannot be nothing so she makes others nothing. This is projecting her nothing outside of herself as a defense. (Anything you disown, don't own, you project outward on another.) She makes others disappear which is called in *Trances People Live* negative hallucination.

Louise: I make people nothing. I then have some relief. I don't have to deal with anything.

Wolinsky: So if you don't know somebody, you make them nothing.

Louise: Yes, I'm afraid all the time that somebody is going to see my nothing. They will see nothing here. They will see I am nothing.

Therapeutic Note

The invisible thing, making others invisible or going invisible yourself is a two edged sword. If you go invisible to survive, you, your needs also go invisible. If I become invisible to mom or dad, I also become invisible to myself. I'm becoming invisible so mom doesn't see me, and then all my needs become invisible to me. Soon my feelings are invisible. My wants are invisible. Many clients that I have seen have I am invisible because my mother was so awful and so I became invisible. I had a client from Michigan and this was her presenting problem. When I'm with my family, "I can't feel my needs," "I can't feel what I want," "I give myself up to my husband and kids and I don't know what I need and I want." So in the process of making herself invisible to mom, *she made herself invisible to herself*.

Wolinsky: How are you feeling now?

Louise: I'm not feeling afraid. I'm not feeling much of anything.

Wolinsky: Have you ever taken on and used this hermit isolationist i-dentity as a way of erasing or rejecting another?

Therapeutic Note

Going outside in from the hermit isolationist (distractor) into the False Core Driver of "I am nothing," "I don't exist." The hermit will a-void and not see people so they are nothing. So they a-void being seen as nothing.

Louise: That's what it is about. I have a fantasy that when someone sees my experience as being nothing that they will reject me.

Wolinsky: Have you ever taken on this hermit isolationist i-dentity as a fear that someone like your mother would engulf you?

Therapeutic Note

Notice how distractors tend to justify themselves (i.e., I must isolate to not feel rejected. Then often it is spirituality, like I must isolate to commune with God, etc. Furthermore this False Core Driver has rejection and engulfment issues. Therefore it is important to include it in the package.

Louise: It is so hard for me to think of *my mother* as engulfing because she *does not exist* but *I was engulfed in that*.

Wolinsky: Say that again. "I was engulfed by my mother's nonexistence."

Louise: "I was engulfed by mother's nonexistence." (The isolation)

Wolinsky: Make the statement, "My mother was nonexistent, therefore I was nonexistent, therefore that's

Chapter X

who I am," and since I was nonexistent to my mother, say, "My mother was nonexistent, therefore I am nonexistent."

Louise: Since I was nonexistent to my mother and my mother was nonexistent, therefore I am nonexistent.

Wolinsky: Did you ever try to match your mother's nonexistence?

Therapeutic Note
People take-on sometimes through "matching" another's state. In N.L.P. or Erickson "matching" and "pacing" is considered a therapeutic tool to gain rapport. However, the "skill" of the therapist to do this "matching" and "pacing" is deeply rooted in the therapist's age regression and survival skills with parents which are duplicated with clients in an unconscious trance-ference loop.

This is detrimental to the therapist's personal growth in that it gives them psychological justificationi and tools to keep their unconscious age-regressioin going. This is an infant's control issue, i.e., the way to survive and maintain control in a situation with mom and dad is to "Take on" and "match" Them. The problem is the "match" (fusion) is forgotten.

This is why these therapies get so into control and are considered by many manipulative. It is because they are fused with an infant's age-regressed manipulation and control issues. *"I wish to apologize to people for my involvement with these systems as a therapist and trainer." At the time, I was unaware of "my" Fusion issues.*

Louise: Yeah.

Wolinsky: Did you forget or pretend that you did that?

Therapeutic Note
We "take on" another False Core and then pretend we did not take it on.

Louise: Yeah. It became an automatic strategy to deal with so much invasion.

Wolinsky: Your mother had to be nonexistent in relationship to your father, so she modeled for you how to be nonexistent with him. Is that correct?

Louise: Uh-huh.

Wolinsky: So intentionally have a father image over there (another part of the room) who's very intense with that force coming at you, and create in response, "I am nothing."

Louise: Okay.

Wolinsky: Good. What decisions did you make around that?

Louise: It was safer to be nothing than fight.

Wolinsky: Is that belief still there?

Louise: Yeah.

Wolinsky: Notice the size and shape of the belief.

Louise: Huge. It's very big.

Wolinsky: Where in your body do you feel the little girl's *I*-dentity? Who did that?

Louise: In my heart.

Wolinsky: Put the little girl i-dentity over there (another part of the room).

Louise: Okay.

Wolinksy:	Notice where in your body you feel "It is safer to be nothing than to fight." Take the label off and let it go back to the little girl *I*-dentity over there.
Louise:	Okay.
Wolinsky:	Now, notice the size and shape of it and the **SPACIOUSNESS** it is floating in and see the **SPACIOUSNESS** and the belief as made of the same substance.
Louise:	Yeah, it's like a hologram from a little girl and father just locked. But I can dissolve the hologram.
Wolinsky:	How are you feeling now?
Louise:	Deeper, quieter.
Wolinsky:	Now, with the father force coming at you this way and you creating in response to the force, the no-force force called "I am nothing,"I have nothing, the invisible, what other decisions did you make?
Louise:	Get away and hide as fast as I could.
Wolinsky:	Where in your body did you hide to get away?

Therapeutic Note
People hide themselves in their body. Therefore we have to locate them so that they are *whole*. Then they can go beyond themselves.

Louise:	It feels like I lost the whole bottom part of my body and then I just hid by bracing, just like making a wall here. I feel some of the physical pain of the wall right now.
Wolinsky:	From holding up the wall?

Louise: Yeah by holding up the wall.

Therapeutic Note

It takes energy to hold up or hold back things.

Wolinsky: You must feel very exposed right now then.

Louise: Yeah, I do.

Wolinsky: Can you say this to me, "I feel very exposed."

Louise: "I feel really exposed right now, Stephen."

Wolinsky: I believe you. I really do. If you took the label off this feeling called exposed and had it as energy, what would that be like?

Louise: I know my own existence and my own energy.

Wolinsky: Can you say that to me? "I know my own existence more, my own energy more."

Louise: "I know my own existence more, and I know my own energy more."

Wolinsky: I want to do one more thing right this second with Dad and that is, we have Dad over there and we have your mother over there. Dad is outrageous so your mother disappears. Correct?

Louise: Uh-huh.

Wolinsky: So what I want you to do is like a vacuum cleaner (metaphor), absorb your mother's state of being nothing, her whole modeling and totally "take on" mother being nothing so she can handle Dad.

Chapter X

Louise: Yeah, I can do that.

Wolinsky: Is that familiar?

Louise: Yes.

Wolinsky: Do it again and report to me any thoughts or decisions that come to mind, if any while you are doing that.

Louise: It's like if I can just hold this long enough, if I can just do this long enough, if I can keep this absorbing long enough, the force will go away.

Wolinsky: What I would like for you to do is flip a switch, blow it out rather than suck it in.

Louise: Oh, she can't handle that. She sure disappears a lot.

Wolinsky: Do you feel like you are nothing right now?

Louise: I've got a lot of body sensations.

Wolinsky: Can you say to me, since I have sensation, then I am and have something.

Louise: (Laughs) I have sensation and I am something.

Wolinsky: Okay. So you do have something.

Louise: Yeah, I do. I have sensations.

Wolinsky: Tell me something else you have.

Louise: A sense of humor.

Wolinsky: Good. Tell me something else you have.

Louise: Emotions.

Wolinsky: How are you feeling now?

Louise: Great. It seems real vibrant. I'm in my body. I can feel my body.

Wolinsky: Say, "I have a body."

Louise: I have a body. I am in it. I have energy in this body too. I have energy right now. It's amazing. I feel dizzy.

Wolinsky: Have you ever taken on and created an observer isolationist *I*-dentity as a way to a-void being engulfed by your mother's nothingness?

Therapeutic Note

The observer is also a defense against another. It is losing awareness of the body to a-void pain.

Louise: Yes.

Wolinsky: Ask the I-dentity to recall a time that you did that.

Louise: I would observe my father and try to help her somehow deal with him. So that she wouldn't do the same things over and over again that would get his criticism.

Wolinsky: Have you ever taken on and used this observer as the only way you could have a separate self, separate from your mother's nothingness?

Chapter X

Therapeutic Note

The observer is an I-dentity and is part of the ego. It is a way to have a separate self.

Louise: Yeah, I could be something with all the information that I got. It gave me an *I*-dentity to have.

Therapeutic Note

Here we see the distraction of "accumulation" of information that the "I don't exist" has so that it knows it exists.

Wolinsky: Have you ever felt special, different or more spiritual after you have taken on this observer isolationist *I*-dentity?

Louise: Special or different? Yeah, and I would tell myself a story that somehow I got it and they don't.

Wolinsky: Have you ever felt more evolved?

Louise: Yes. (Laughs) My feet hardly touch the ground. I wanted to be a saint. A Catholic saint.

Wolinsky: Have you ever taken on this observer *I*-dentity and withdrawn as a way of punishing or pusing another away?

Louise: Yes.

Wolinsky: *Prior* to taking on this I'm nothing observer thing, was there any chaos or shock that occurred?

Louise: I was born. It just feels like it was always there. The image that just came up with is that baby in 2001 floating in space. But that feels like what it was. I disappeared as a baby.

Therapeutic Note

This False Core is the earliest False Core Driver solidified. It is solidified in-utero because it is in-utero where rejection and the separation are experienced. This interest in being surrounded by space"dissociative void" also explains the draw of this False Core into Buddhism and the "space" thing. It should be noted that there is a difference between space which is a dimension (Einstein's space-time) and the **BIG NOTHINGNESS** (see Volume III).

Wolinsky: Say that again.

Louise: Stephen, "I disappeared as a baby." "I disappeared as a baby."

Wolinsky: And do you have the image of the baby?

Louise: Yes.

Wolinsky: Is it a baby before or after birth?

Louise: Before.

Wolinsky: So notice the image of the baby before birth in the womb. I would like you to do two things. First I want you to merge with the baby in the womb.

Louise: I really have got to contract, okay.

Wolinsky: And now, as the baby in the womb rather than breathing, experience yourself *being breathed*.

Louise: There's hardly anything there. There's hardly anything.

Wolinsky: Can you say to me, "I'm hardly anything?"

Louise: "I am hardly anything."

Chapter X

Wolinsky: What's happening with your breathing now?

Louise: The breathing is happening.

Wolinsky: Good.

Louise: In there I had to breath just a little to not disturb anyone or not be noticed.

Wolinsky: So that?

Louise: So that I could exist and not disappear.

Wolinsky: And not?

Louise: Disappear. I did that. I breathed just a little so I could exist. I did that.

Wolinsky: I want to give you one little phrase between that. I breathed just a little so I wouldn't take up too much space. So then I could exist.

Louise: I breathed just a little so I wouldn't take up to much space. So then I could exist. I then could exist. I breathed a little so I wouldn't completely take up the little room that was there so I could exist.

Wolinsky: Yes. So what did you decide after that?

Louise: That's the way life was.

Wolinsky: Are you still deciding that right now?

Louise: No. But I can see the chain of associations.

Wolinsky: How are you feeling right now?

Louise:	I had an extremely strong sense of rejection in the womb from my mother.

Therapeutic Note

That is the "I don't exist" distractor. This False Core Driver has that schizoid tendency. It is developed in-utero, (Alexander Lowen) by the rejection from the mother.

Louise:	I have been telling myself all of these years that my mother did want me. That's the story. I realized that who told me that she wanted me was not mother, but my aunt. I bought the lie. So I was pretending my mother wanted me. I spent all of these years pretending that she wanted me. Then, I had to pretend my father did want me, so with my father I had to pretend that I existed. That was the deal.
Wolinsky:	How does all this seem to you now?
Louise:	Very distant.
Wolinsky:	How do you feel now?
Louise:	Free, beyond my associations—very spacious.

QUANTUM PSYCHOLOGY PRINCIPLE:

Meaning should be a function of context. Meaning is a function of the level of abstraction.

CONCLUSION

Someone asked Nisargadatta Maharaj, "If you are enlightened, What's the weather like in Cincinnati?" He would say, "Whatever I need to know, I will know, whatever I don't need to know, I won't know." He took all the charge off of having to know. The #1 Distractor in this False Core-False Self is "I don't know-I have to know." The False Core-False Self is not about pure knowing (i.e., knowing with no subject-object. Pure knowingness is just pure knowingness. *Knowing about* is the perception of two or more substances. Like *I know about* how to fix a fence. It has nothing to do with *pure knowing with no subject-object.* People sometimes confuse being psychic (knowing about) with pure knowing with no subject-object.

Knowing about distracts one from the "I do not exist" False Core. Trying to know (about) is the False Self's solution to the False conclusion.

EXERCISE

1. Without using your thoughts, memory, emotions, associations or perceptions, Do you exist, not exist or neither?
2. Without using your thoughts, memory, emotions, associations or perceptions, what does existence or non-existence ever mean?
3. Without using your thoughts, memory, emotions, associations or perceptions, what does empty mean?

***THE NON-VERBAL I AM*: PURE KNOWING IN QUANTUM PSYCHOLOGY MEANS NO SUBJECT-OBJECT, "NO FRAMES OF REFERENCE, NO REFERENCE TO FRAME."**

Chapter X

It is the False CORE which labels itself as a sinner or full of vice. It is the *False Self* which attempts to change, transform or convert vices into virtues.

This subtle, seductive tendency is laden with judgments, evaluations and significance like lazy, fear, lust, anger, pride, envy, deceit, avarice and gluttony are vices and sobriety, innocence, right action, courage, honesty, humility, serenity, balance, nonattachment are virtues. Unfortunately, these judgments about bad and good are oftentimes introjected parents or societal. Since they are False Core-False Self concocted, any attempt to convert vices into virtues ultimately re-enforces the False Core.

CHAPTER XI
THE WAY OF THE LONER

FALSE CORE DRIVER:
I am Alone.

FALSE SELF COMPENSATOR:
I must not be alone—I must connect, the over-connector.

The loner is the outcome of this False Core Driver whose conclusion is "I am alone." As with all other False Cores, this one sets itself up to re-enforce and prove that it is alone. As with all False Self Compensators, this False Self Compensator will adapt "spiritual and psychological philosophies" to *try* to compensate for or justify the virtues of being alone (that it's more spiritual, etc.). The greatest fear of this False Core, more than being alone, is to be shunned. This False Core Driver has the False Self Compensator and obsessive-compulsive tendency to *"over* connect" to overcome the "I am alone."

As with all False Self Compensators, the *over* is false. At the time of connection there is a " high" and a relief from "I am alone." However, like a drug addict, the False Self Compensator needs more and more of the compensation to get the same"high," the same relief. This is how it works with a spiritual or psychological practice or technique which is used to help make a "connection." During moments of spirituality, meditation, or psychological contact or insight there is a relief from the False Core but as soon as you stop the pain comes back. "Spiritual and Psychological" disciplines are often unaware of the False Core-False Self phenomenas and are unable to handle the underlying pain. Thus, they might suggest to students and disciples in pain to just do the spiritual practice *more*—do more mantra, yantra, tantras or communicate, feel your feelings, heal, etc. and the pain will go away. But this only *re-enforces* the False Core Driver's pain (doing it wrong) and the False Self Compensator's doing it more since *it is the False Self Compensator who is doing the spiritual or psychological practice.* In other words, the False Self Compensator is seeking Nirvana (connection with Mom) again.

QUANTUM PSYCHOLOGY PRINCIPLE:

Psychological and Spiritual Disciplines are often done by the False Self Compensator. This is why it is difficult to stabilize or go beyond pain. The False Self-False Core is a holographic unit and the False Self must always keep the False Core present and in its awareness so it knows what it is working on.

Please note that all False Core Drivers-False Self Compensators can use substitutes so as "not to feel alone," for example, connecting with a thought, a feeling, nature, the past, the future, etc. All False Self Compensators are like a drug. In this case, any connection (drug) will do. Of course, each False Core Driver-False Self Compensator has its drug of choice. To illustrate:

FALSE CORE DRIVER -
FALSE SELF COMPENSATORS

The False Core "I am *imperfect*" may seek (False Self) perfection through cleanliness.

The "I am *worthless* False Core " might seek (False Self) worth through getting flattered.1

The "I am inadequate" False Core might seek (False Self) merger through reasonableness.

Chapter XI

The "I am *non-existent*" False Core might seek (False Self) existence through spiritual immortality

The "I am *alone*" False Core might seek (False Self) connection through work.

The "I am *incomplete*" False Core might seek (False Self) merger through experiences.

The "I am *powerless*" might seek (False Self) merger through power.

The "I am *loveless*" False Core might seek (False Self) merger through love and a New Age Christianity.

THE REASON FOR I-DENTITIES

I am alone because (*fill in the blank*) (I am unacceptable, I am a burden, I am empty).

FALSE CORE DISTRACTORS

People with this False Core distract themselves from being alone by setting themselves to be weak and exhausted.
To defend against the weak which distracts them from "alone," they put a lot of time and energy into becoming overly strong.

Emotional Distractor:
Fear, Paranoia, Self-Doubt and Terror. Simply put, fear is the main distractor of attention so the False Core does not have to be experienced, known or felt.

Spiritual Defenses:
Spiritualizing, "I am alone," a monk, swami, hermit, unmarried yogi type.

Ichazo *I*-dentities
I-dentity #1 struggle	or	*I*-dentity #1 pushy
I-dentity #2 give up	or	*I*-dentity #2 surrender

Quantum Psychology
False Core Distractor—weak
False Core Distractor—strong

IS THIS A FEAR THING?

Quantum Psychology sees the fear of this False Core as a distraction of attention away from the False Core and as a reaction to the weakness of the infant who has just gone through the shock of the Realization of Separation and its conclusion "I Am Alone and do not want to be shunned." In other words, I would rather feel fear than feel alone and shunned. For example, they (the infant) can imagine themselves in the future where something they are"too weak" to handle what might happen. They feel fear and do not deal with the "I am *Alone*." Or they could imagine something happening and then get *pre*-occupied with being strong and being able to handle it—all to a-void I am alone.

THE PROTOCOLS:
DISMANTLING THE FALSE CORE-FALSE SELF

False Core Driver—"I Am Alone"

1. Where in your body do you experience this *"I am alone" I-dentity*? Write down the answer until nothing pops up.

2. Who if anybody modeled this *"I am alone" I-dentity* for you? Write down the answer until nothing pops up.

Therapeutic Note

At this point it is advantageous to externalize the *I*-dentity. This can be done in the following way:

Step I: Notice where in the present time body the *I*-dentity is located.

Chapter XI

> Step II: Notice the *I*-dentity's size and shape.
> Step III: Take the label off of the *I*-dentity. Have it as energy.
> Step IV: Allow the *I*-dentity as energy to move from the present time body to another physical location in the room, and then to become solid again.

3. Regarding physical illnesses, what did your mom or dad tell the *I*-dentity about members of the family, or were they always in a kind of a panic state, warning you about what might happen if you didn't take care of a scratch (e.g., that it would get an infection). Ask the *I*-dentity this question and write down the answer until nothing pops up.

4. What did others tell the *I*-dentity (suggestions) that were given to you regarding *"I am alone?"* Ask the *I*-dentity this question and write down the answer until nothing pops up.

5. This *"I am alone" I-dentity* and empty label, did it ever feel fear to distract itself from *I am alone*? Ask the *I*-dentity this question and write down the answer until nothing pops up.

6. Has this *"I am alone" I-dentity* ever been used to create psychosomatic illnesses to help you feel and justify being alone? Ask the *I*-dentity this question and write down the answer until nothing pops up.

7. Has this *"I am alone" I-dentity* ever experienced what you would think are illnesses which really, when you explore them, repressed fear to hide *"I am alone?"* Ask the *I*-dentity this question and write down the answer until nothing pops up.

8. Has the *"I am alone" I-dentity* ever used weakness or fear and then immediately changed it into anger? Ask the *I*-dentity this question and write down the answer until nothing pops up.

9. Has this *I*-dentity of *"I am alone"* ever felt victimized and tried to persecute another? Ask the *I*-dentity this question and write down the answer until nothing pops up.

10. Has this *"I am alone"* *I*-dentity ever been used as a spiritualized *I*-dentity, i.e., "I surrender" or "Whatever happens happens, it's up to God," "Giving up to a higher power," etc. Ask the *I*-dentity this question and write down the answer until nothing pops up.

11. Has this *"I am alone"* *I*-dentity ever been used to justify being alone, unconnected, not in a relationship, even if you were? Ask the *I*-dentity this question and write down the answer.

12. Has this *"I am alone"* *I*-dentity ever been used to justify the importance of aloneness as a spiritual practice. Ask the *I*-dentity this question and write down the answer.

13. Ask the *I*-dentity to recall a time at an early age when it felt alone or weak. Ask the *I*-dentity this question and write down the answer until nothing pops up.

14. Ask the *I*-dentity to recall a time at an early age when it felt the physical demands were too much and that it still had to do it. Ask the *I*-dentity this question and write down the answer until nothing pops up.

15. Ask the *I*-dentity to recall how it might have reacted as a rebel when in school, particularly in first grade to resist the feelings of *"I am alone."* Ask the *I*-dentity this question and write down the answer until nothing pops up.

16. If this *"I am alone"* *I*-dentity were no longer there, what would the experience be? Ask the *I*-dentity this question and write down the answer until nothing pops up.

Chapter XI

17. Has this *"I am alone"* I-dentity ever been used to give an air of spirituality and then because it was so weak you felt indifferent? Ask the *I*-dentity this question and write down the answer until nothing pops up.

18. Has this *"I am alone"* I-dentity ever been used as a way of imagining it was in a higher level of spiritual understanding? Ask the *I*-dentity this question and write down the answer until nothing pops up.

19. Has this *"I am alone"* I-dentity ever been used to decide that life was meaningless, which you then believed was a higher level of spirituality? Ask the *I*-dentity this question and write down the answer until nothing pops up.

20. Has this *"I am alone"* I-dentity ever been used to create a feeling of being special, different or more spirituality or psychologically evolved than others? Ask the *I*-dentity this question and write down the answer until nothing pops up.

21. Has this *"I am alone"* I-dentity ever been used as a way of justifying actions? Ask the *I*-dentity this question and write down the answer until nothing pops up.

22. During the *I*-dentification with this *"I am alone"* I-dentity and empty label, was there any chaos or shock that occurred? Ask the *I*-dentity this question and write down the answer until nothing pops up.

23. Regarding this *"I am alone"* I-dentity and empty label, what decisions, beliefs, or assumptions did it make? Ask the *I*-dentity this question and write down the answer until nothing pops up.

24. During the process of taking on this *"I am alone"* I-dentity it did ever have any fantasies of an early death? Ask the *I*-dentity this question and write down the answer until nothing pops up.

25. Has this *"I am alone" I-dentity* ever been used to justify excess needs of rest and isolation? Ask the *I*-dentity this question and write down the answer until nothing pops up.

26. Notice where in your body you experience this *"I am alone" I-dentity*?

27. Peel back the *"I am alone" I-dentity* and the empty label (if it is there) and notice the **SPACIOUSNESS** underneath it.

28. From inside the **SPACIOUSNESS**, ask the *"I am alone" I-dentity* and empty label, what are you seeking more than anything else in the world, and allow them to answer. Write down the answer until nothing pops up.

29. Experience the essential quality that the *I*-dentity and label is seeking inside the **SPACIOUSNESS**, and as you continue to feel that, notice the size and shape of the *"I am alone" I-dentity* and the **SPACIOUSNESS** that surrounds it, and see the **SPACIOUSNESS** and the *I*-dentity as being made of the same substance as you continue to experience the essential qualities.

FALSE SELF COMPENSATOR: "I HAVE TO CONNECT."

1. This *"I have to connect" I-dentity*, where do you feel it in your body? Ask the *I*-dentity this question and write down the answer until nothing pops up.

2. Allow yourself to experience the *"I have to connect" I-dentity*.

3. Now experience the *"I am alone" I-dentity* in your body.

4. Allow yourself to slowly move from the *"I am alone" I-dentity* into the *"I have to connect" I-dentity* and notice if the *"I have to connect" I-dentity* over-compensates for the *"I am alone" I-dentity*. Write down the answer.

Chapter XI

5. Has this *"I have to connect" I-dentity* ever been used as a way of overcoming, *"I am alone?"* Ask the *I*-dentity this question and write down the answer until nothing pops up.

6. Has this *"I have to connect" I-dentity* ever been used as a way of pushing through weakness, illness, or tiredness? Ask the *I*-dentity this question and write down the answer until nothing pops up.

7. Has this *"I have to connect" I-dentity* ever been used as a way of pushing through emotional, physical, sexual, or mental weakness? Ask the *I*-dentity this question and write down the answer until nothing pops up.

8. Has this *"I have to connect" I-dentity* ever been used in an attempt to beat the system, because you felt beaten down by the system? Ask the *I*-dentity this question and write down the answer until nothing pops up.

9. Has this *"I have to connect" I-dentity* ever acted as a rebel as a way to connect and a-***void*** "I am alone?" or to set yourself to be alone and justify it. Ask the *I*-dentity this question and write down the answers until nothing pops up.

10. Has this *"I have to connect" I-dentity* ever created itself as an authority figure? Ask the *I*-dentity this question and write down the answer until nothing pops up.

11. Has this *I*-dentity ever used work as the only or best way and justified it to *connect* with others and overcome your *"I am alone,"* and then justified it in some way? Ask the *I*-dentity this question and write down the answer until nothing pops up.

12. Has this *"I have to connect" I-dentity* ever been used as way of looking down on people who could not connect, or who were weak (physically, emotionally, sexually, spiritually, mentally)?

Ask the *I*-dentity this question and write down the answer until nothing pops up.

13. Has this *"I have to connect" I-dentity* ever been used to look down on people who gave up? Ask the *I*-dentity this question and write down the answer until nothing pops up.

14. Has this *I*-dentity ever felt strong for not showing your needs? Ask the *I*-dentity this question and write down the answer until nothing pops up.?

15. Has this *"I have to connect" I-dentity* ever decided that feeling or having needs represented weakness and, therefore, was bad? Ask the *I*-dentity this question and write down the answer until nothing pops up.

16. Has this *I*-dentity ever imagined having no needs was the best way to connect? Ask the *I*-dentity this question and write down the answer until nothing pops up.

17. Has this *I*-dentity ever been attracted to a spiritual system where needs were looked on as bad or weak or unevolved? Ask the *I*-dentity this question and write down the answer until nothing pops up.

18. Has this *"I have to connect" I-dentity*, ever bought into a spiritual system that labeled desires or wants as weak, worldly, or bad and that the *way to connect* (i.e., merge) was through giving up and not having needs? Ask the *I*-dentity this question and write down the answer until nothing pops up.

19. Has this *"I have to connect" I-dentity* ever been used as a way to justify and spiritualize life's struggles as a way to merge-connect? Ask the *I*-dentity this question and write down the answer until nothing pops up.

Chapter XI

20. Has this *"I have to connect" I-dentity* ever become the strong rebel as a way to connect? Ask the *I*-dentity this question and write down the answer until nothing pops up.

21. Has the *"I have to connect" I-dentity* ever become a strong rebel who projects weakness on others and is going to champion people who couldn't take care of themselves like the weak, under-privileged, and downtrodden people? Ask the *I*-dentity this question and write down the answer until nothing pops up.

22. Has the *"I have to connect" I-dentity* ever used persuading others so they will reflect its opinion back and it will feel connected? Ask the *I*-dentity this question and write down the answer until nothing pops up.

23. Has the *"I have to connect" I-dentity* ever felt special, different or more spiritual or psychologically evolved from others? Ask the *I*-dentity this question and write down the answer until nothing pops up.

24. During the emergence→*I*-dentification with the *"I have to connect" I-dentity*, was there any chaos or shock that occurred? Ask the *I*-dentity this question and write down the answer until nothing pops up.

25. Has the *"I have to connect" I-dentity* ever been used to cover up feelings of fear and being a coward? Ask the *I*-dentity this question and write down the answer until nothing pops up.

26. During the process of *I*-dentification with this *"I have to connect" I-dentity*, what assumptions, decisions or beliefs did it organize around? Ask the *I*-dentity this question and write down the answer until nothing pops up.

27. During the process of this emergence→*I*-dentification of this *"I have to connect" I-dentity*, were there any fantasies about being superhuman, chosen, super-spiritual, or acknowledged

in some way as somebody super-strong, super-intelligent, super-spiritual, super-sexual, super-physical, etc.? Ask the *I*-dentity this question and write down the answer until nothing pops up.

28. Notice where in your body you experience this *"I have to connect" I-dentity*. Write down the answer until nothing pops up.

29. Peel it back and the empty label underneath that, and notice the *"I am alone" I-dentity* underneath it.

30. Peel these back and notice the **SPACIOUSNESS** and go into the **SPACIOUSNESS**.

31. From inside the **SPACIOUSNESS**, notice the two *I*-dentities and ask them, "What are you seeking more than anything else in the world?" (allow them to answer). Write down the answer until nothing pops up.

32. Experience in the **SPACIOUSNESS** of **ESSENCE** the essential qualities that the two *I*-dentities and the empty label seeking.

33. Notice the size and shape of those two *I*-dentities and the empty label and the **SPACIOUSNESS** they are floating in. See the **SPACIOUSNESS** and the *I*-dentities and the label as being made of the same substance, while you continue to experience the essential qualities inside the **SPACIOUSNESS** of your own **ESSENCE**

Demonstration

Deborah is a psychotherapist about 60 from Chicago. With Deborah, we first focused on her *weakness* distractor. Oftentimes we start with the distractor which leads to the False Core.

Wolinsky: Where do you feel the "I am alone" *I*-dentity in your body?

Deborah:	At this moment. Probably in my eyes, back of my throat.
Wolinsky:	So go way behind your eyes. Do you feel it?
Deborah:	Yeah.
Wolinsky:	How does the rest of your body feel as you go in the weak/"I am alone" *I*-dentity?
Deborah:	Again, I feel a slackening of the tension of the jaw and mouth. Well I feel a curl in the shoulders but yeah the heart falls, droops, pulls in. A kind of drooping heaviness settles in.
Wolinsky:	So, intentionally droop.
Deborah:	It's a real shift of energy. It's a major shift. It's almost like I'm pulling my energy right inside.
Wolinsky:	Now, I would like for you to have this image of your mother over there (another part of the room).
Deborah:	Okay.
Wolinsky:	Feel the weak "I am alone" *I*-dentity. Now, look at your mother. I would like for you to be in the weak "I am alone" *I*-dentity and very slowly use your awareness and organize and become the "I have to connect strong" *I*-dentity in your mother's presence.

Therapeutic Note

In this demonstration, we are going for the distractor *I*-dentity weak/strong and the False Core Driver-False Self Compensator of "I am alone, I have to connect." So we ask her to automatically create strong (since that is what she is already doing) whenever weak comes up. I assumed it had something to do with one of her parents.

Deborah: There is tightness. It straightens. Again, behind the eyes there's a total shift from the place that I experience myself as being shifts behind the eyes so that when I am looking at her, shifting from the weak I am alone to strong I have to connect, I can almost feel an intense kind of defiance.

Therapeutic Note

Always get the *external context* this is why I use the images of mom, etc. Later the trance-ferences of mom on another is the next step which triggers the internal response, thus, the whole loop will become revealed.

Wolinsky: Is strength also a defiance?

Deborah: Yes. It's like I *will not*.

Wolinsky: So now looking at your mother again here, shrink back into the weak, I am alone.

Deborah: Okay.

Wolinsky: Okay, look at me and make a statement. It hurts to be that separate.

Deborah: It hurts to be that separate.

Wolinsky: Now, looking at your mother, just make the statement to her, it hurts to be separate.

Deborah: It hurts to be separate.

Wolinsky: Is that true?

Deborah:	Very much so.
Wolinsky:	How about "I won't show you my separateness."
Deborah:	I won't show you my separateness. It becomes strong, like I don't need anyone; I have no needs.
Wolinsky:	Is that part of defending against being separate?
Deborah:	Whoa. I feel like running away.
Wolinsky:	Is it is okay to be weak and alone now in this relationship?
Deborah:	Yeah.
Wolinsky:	What happens to your body?
Deborah:	Well, my body doesn't even take her in. I feel the separateness. It's like it's not there. The thing that I like. Stephen wait a minute, I'm going to cry. I think that's another *I*-dentity. It's the one that wants the connection. I don't know what to do with that in between. It's sitting here. I slid into that wanting.
Wolinsky:	So, what I would like for you to do is to go into the separate, alone, weak *I*-dentity. Look at your mother for a second. And identifying with the *separate/weak I-dentity*, feel the wanting. Got that. Now from that wanting, since you did not get what you wanted, what did you assume, decide or believe about not getting what you wanted?
Deborah:	*"I would not need."*

Therapeutic Note

Here needing is fused with the shock of the Realization of Separation.

Wolinsky: So with the hurt, can you say it hurts to feel separate?

Deborah: Yeah. "It hurts to feel separate."

Wolinsky: Now, can you say, "I don't want to feel separate so I will feel defiant, strong and have no needs."

Deborah: "I don't want to feel separate so I feel defiant, strong. Yes, I have no needs."

Wolinsky: Is that true?

Deborah: Yes, absolutely.

Wolinsky: Since you said, you decided, that no matter what, you would not be alone, what did you decide to do when aloneness came up for you?

Deborah: I filled it up with things.

Wolinsky: Such as?

Deborah: Activities, and people and too much work and things that kept me focused on other things and I try to connect through them, but it never works.

Wolinsky: How are you doing now?

Deborah: Well, part of me is in that space where I created this and do this, and the other half of me is not, which is very nice.

Wolinsky: Your puffiness around your eyes has gone down very suddenly.

Deborah: Really.

Wolinsky: Have you ever taken on this *I*-dentity and used it as a "weak alone" *I*-dentity, or as a spiritual *I*-dentity?

Deborah: I spiritualize the hands. Like these hands have the capacity to do healing.

Wolinsky: Have you ever taken on this *I*-dentity and used it as a spiritual *I*-dentity as a way of surrender or as way to connect, like through healing work?

Deborah: Yes, exactly, so my weak hands then became strong. Then, I connect with others.

Wolinsky: Okay. Now they are strong, healing hands that can connect through healing work.

Deborah: Oh, yes.

Wolinsky: Tell me another thing that you did that you felt too weak to do.

Deborah: You know it is interesting, as you are saying this, I am focusing in on how I split my body in half, literally. I have strong bottom, legs and feet but my hands and arms were weak. That's where I would hold this weakness was in my hands and arms and could feel the strength.

Wolinsky: Tell me another thing that you were asked to do that you felt too weak to do.

Deborah:	I am still going to go back to another weak time. I was told and expected to be taken away from my family and not need anything about that. I feel bad, made a fuss.
Wolinsky:	You repressed your loneliness and focused on no-needs or strength and connection. Where did you put aloneness, no need, and weak in your body?

Therapeutic Note

Once again, weak is the distractor for aloneness. But she will not feel the weak. Instead she becomes overly-strong with no needs as a way to connect with others.

Deborah:	In my gut.
Wolinsky:	Is that where the split is?
Deborah:	Yeah, exactly, on top of it.
Wolinsky:	How do you feel now?
Deborah:	I actually have a feeling of heaviness like my head and part of my body feels—I feel lighter here. I'm not together. I'm numb.
Wolinsky:	So you feel disorganized.
Deborah:	Yeah. It's not totally comfortable. My tongue doesn't even feel like it wants to work right. I have numbness of the tongue. That's interesting.
Wolinsky:	So tell me a time you had to numb your tongue so that the needs or what you really wanted to say couldn't come out, or emotionally you could not express because your tongue was numb.

Chapter XI

Therapeutic Note

Her tongue goes into the numbness trance (sensory distortion to avoid saying what she feels, wants, etc.)

Deborah: A lot of times, family. You would tell them the truth and there wasn't any ears for that.

Wolinsky: Tell me another time.

Deborah: You bite it off. You don't say your truth.

Wolinsky: How does your tongue feel now?

Deborah: Better.

Wolinsky: Where do you feel the weak, kind of exhaustion?

Deborah: Well I go collapsing all over.

Wolinsky: How are you doing?

Deborah: It's nice to feel collapsed.

Wolinsky: Yeah.

Deborah: I almost feel like celebrating because I'm staying here. It doesn't hurt.

Wolinsky: It hurts to resist the collapse.

Therapeutic Note

It does not cause pain to feel the collapse, just the resistance to the collapse.

Deborah: I just realized I repress my aloneness along with my sexual needs.

Wolinsky: Say that again.

Deborah: I repress my feelings of aloneness along with my sexual needs. So I don't have to need or show my needs to another.

Therapeutic Note:
In this context (1 hour) I do not do—but I could and would, in another session, a six-step de-fusion process of alone = no needs (sexual).

Wolinsky: How does it seem to you now?

Deborah: Well, I can see part of that defiance in response to my mother. It's a defiance. The other side of it.

Wolinsky: The other side of it means what?

Deborah: Well, it was like it was here and I was talking about it and connecting with it. It feels like even that the saying of it has shifted it so that it is not—my relationship is feeling different. I'm not as attached to it somehow and my gut hurts.

Wolinsky: Describe the hurt to me.

Deborah: It is like a band of fire across my upper belly, my upper diaphragm.

Wolinsky: Take the label off the band of fire. What happened?

Deborah: It's energy.

Wolinsky: How are you feeling?

Deborah: Hot.

Wolinsky: How does that split seem to you now?

Chapter XI

Deborah: Well, the split isn't where it was.

Wolinsky: Is it there at all? How are you feeling now?

Deborah: Cool.

Wolinsky: Is that cool?

Deborah: Well, I'm warm. My butt is hot.

Wolinsky: So, how does that belief, "If I repress my aloneness and my sexuality, then I don't need anybody." How does that seem to you now?

Therapeutic Note

The False Core is fused with many things, feelings, thoughts, and in this case, sexuality. To see and be able to have the False Core as itself, it must be as it is (just a concept) with no fusions or associations first.

Wolinsky: Now, notice the weakness, no needs, "I am alone," and be just prior it, go into your **ESSENCE** (the **SPACIOUSNESS** inside your body). And how does the weak, no needs, "I am alone" and the strong connected seem to you from inside your own **ESSENCE** now?

Deborah: It's nothing. They aren't there.

Wolinsky: So what is it that has thousands of eyes and can't see?

Deborah: Silence (and a nod). I feel wonderful. My body is different and spacious.

DEMONSTRATION— UNFUSING THE FALSE CORE FROM THE ENTIRE PSYCHOLOGY

The False Core is the puller of the chain of associations which is your psychology. You must dismantle the False Core from its association otherwise, the False Core will continue to create associations.

QUANTUM PSYCHOLOGY PRINCIPLE:

All associations lead to the False Core and, if not dismantled, the associations will only re-enforse and justify themselves and the False Core.

QUANTUM PSYCHOLOGY PRINCIPLE:

Any attempt to change, reframe, reassociate or re-decide associations only adds to the number of associations (links in the chain) and serves to strengthen the chain, making it more difficult to dismantle the False Core.

DEMONSTRATION:
DEFUSING THE FALSE CORE FROM THE ASSOCIATIONAL CHAINS.

Wolinsky: If the concept of alone is fused with your entire psychological, emotional and biological life, what gets created?

Bill: Separateness.

Wolinsky: What did this alone concept assume, decide, or believe that got to creating separateness?

Bill: That it's separate.

Chapter XI

Wolinsky: Are you believing it?

Bill: No.

Wolinsky: Now, if you are separate from the "alone concept" and from all of its[1] associations and psychology, what is the "alone concept" not creating?

Bill: All this stuff to resist that and continue. To overcome it. To try to fix it. That's what all the rest of psychology is creating if it is attached to the False Core. If it is separate from the alone concept, there is only space.

CONCLUSION
1. Without using your thoughts, memory, emotions, associations or perceptions, are you alone, connected or neither?
2. Without using your thoughts, memory, emotions, associations or perceptions, what does alone, connected or shunned even mean?
3. Without using your thoughts, memory, emotions, associations or perceptions, what does empty mean?

**NOTICE THE NO-STATE
STATE OF THE NON-VERBAL I AM.**

---

[1]Please note the False Core or "it" has associations or a psychology you do not, (see Volume III).

CHAPTER XII
THE WAY OF THE INCOMPLETE

FALSE CORE DRIVER:
"I am incomplete," "There must be something missing," "I am not enough."

FALSE SELF COMPENSATOR:
I must get whole, complete, completed or full through experiences.

The False Core "I am incomplete" or "There must be something missing" seeks their imagined Nirvana (merger with mom) through experiences; hence, they seek and become over-experiencers. The deep incompleteness of this False Core yields an obsessive-compulsive tendency to seek experiences to a-void the shock of Separation and to somehow "fill in what is missing or where they feel not enough." In other words, because of the shock, they concluded, "I am incomplete, there must be something missing, I am not enough." If only I can have and get more and more, or find out what you have got or get what you got, I could fill in the missing pieces and feel complete." If only this "missing piece" within me were filled I would be whole and then I would be able to reach Nirvana enlightenment (oneness with mom).

As with all False Cores, the False Self Compensator's quest for the Holy Grail is never enough since as with all "highs," you come down into your False Core yet again. In the case of this "pretend happy" and "fake optimistic" False Self, there is a major dissociation between the pain of the False Core Driver and The False Self Compensator. This means that the pleasure seeking of this False Self, knows nothing of the pain of incompleteneess which is driving its pleasure seeking. In other words, as with all False Core-False Selfs, there is a layer of amnesia which separates the False Core from the False Self. *In this False Core-False Self complex, there lies the deepest strongest and most powrful amnesia of all.*

They are pleasure seeking to feel complete and are unknowingly on a treadmill. Once the treadmill stops, if even for a moment, they must find yet another positive, pleasurable experience to overcome the incomplete one. It is extremely difficult for them to get in touch with their pain or allow another person to have their pain. They are *driven* to constantly see the "bright side" otherwise their own pain will get re-activated.

Spiritual Defense:
Using spirituality and seeking blissful "heart experiences," or reframing pain into lessons, solutions, motivations or opportunities to overcome the pain of the False Core. This False Self *unknowingly* creates spiritual rules and standards which can never be met in spiritual teachers or teachings. In this way, they never have to be confronted with their imagined incompleteness, but rather run from teacher to teacher (experience to experience), never going deep, but rather gathering "spiritual" experiences to overcome their False Core which can never work. I've known many people with this False Core-False Self who distracted themselves and lived by the super-standards and rules of a rigid[1] spiritual discipline" as if" this would bring them to God (Mom) and away from "I am Incomplete."

---

[1] Using Ichazo's rigid I-dentity to handle the no-rules of "I am incomplete"

Chapter XII

PSYCHOLOGICAL DEFENSE: REFRAMING AND SOLUTION-FOCUSED THERAPY

The puer or Peter Pan of the False Core-False Selves whose obsessive-compulsive drive to seek Nirvana (merger with mom) and a-void pain leads them into their age regressed denial of seeking a positive experience in everything. For this reason, reframing techniques would be of little help because these approaches reframe and redefine experience as being "positive." This False Self is on automatic reframe, thus unknowingly re-enforcing the False Core through resistance. New Age Christianity sometimes uses this False Self as it looks for "the positive" lesson within every experience. This, of course, re-enforces and perpetuates this False Core compensator.

I once met a False Core-False Self who worked with people trying to help them find "positive solutions" to their problems. This kind of work only served to re-enforce his False Core-False Self cycle as he and his clients continued resisting the pain of their False Core.

For this False Core of "*I must experience*" *anything but incompleteness*, the False Self compensator's defense of the over-experiencer can brag and create the illusion of being successful as long as they keep moving rapidly from experiencing to fantasizing to planning thus a-voiding their False Core of "I am incomplete," "There must be something missing."

THE REASONS FOR I-DENTITIES

I am incomplete because there is something missing, I am empty (*fill in the blank*).

FALSE CORE DISTRACTORS

This False Core-False Self *unconsciously* distracts itself with the idea that there is no wisdom in the world and no rules. People with this distractor seem whimsical and flippantly seek pleasure. Since there are no rules, they will do anything to a-void the pain of "I am incomplete." This superficiality, however, hides the deep pain of feeling incomplete. To defend against this, they become planners (Ichazo), always planning for the future. This constant futurizing likes to do, see, have, create, plan for a nest egg or enlightenment as ways to fill-up the empty feeling of "Something is missing." In other words, after I have this or that experience, I will be complete. This manifests as planning for future experiences so that then I will be complete. Note the process is planning for (*fill in the blank*). They just want to have a good time to fill up this incomplete feeling.

Planning can also be seen as an attempt to have a future where they see none. (See Volume III.)

The problem is that there is a deep dissociation and age-regression from the external. They are trying to compensate, overcome, reframe problems into solutions, heal or transform (out of the mind of an infant) in present time to make up for what occurred in past time. This delusional tendency can be seen in many therapies which maintain that:

1. The therapist can give you the corrective experience you did not have.
2. You can re-program or reassociate the "past."
3. You can create the future in the present.

In four-dimensional space-time, however, the past, present and future are as they are (to be discussed in Volume III).

QUANTUM PSYCHOLOGY PRINCIPLE:

You cannot overcome, one dimension with another no matter how hard you try or how creative you are.

Ichazo's I-dentities
I-dentity #1—superior
I-dentity #2—inferior.

Because of the shock of the Realization of Separation, they believe there is *no wisdom*. Then, as a reaction formation to that conclusion they become idealists. For this reason, Ichazo calls them over-idealistic and states that they create hierarchies.

The False Core-False Self creates and destroys rules simultaneously rather than being free to have or not have rules depending upon the external context. They plan for the future as a way of constructing a solid universe. They have "There are no standards or rules" I-dentity and an automatic over-compensation for *no standards or rules* with super-standards and rules for people, spiritual systems, and spiritual teachers. Naturally, the "teachers" cannot meet their unconscious super-standards, because if they did they would have to feel I am incomplete etc. This gives them a feeling of control.

QUANTUM PSYCHOLOGY PRINCIPLE:

There is no wisdom in the external world, but *there is essential wisdom,* but *not* as a wisdom *about* but as *pure*—with no subject—object.

In the *Tao of Chaos* I called this, "I'm okay if I fill you up, you're okay if you fill me up." There is a constant desire to fill up. Ichazo calls them gluttons—gluttons emotionally, gluttons trying to have experiences, etc. They also seek completeness by trying to raise their status by being associated with and turning people onto people who they see as higher up economically, spiritually, etc., and then trying to bring them down, or put them down, so that the incomplete False Core does not feel as incomplete and can feel more complete. In short, if they put you on a pedestal and then bring you down, they feel more complete or less incomplete. In this way, they try to feel full and complete to a-void the "missing" feeling in their belly or in their mind, thus Ichazo's superior/inferior *I*-dentities.

PROTOCOLS:
DISMANTLING THE FALSE CORE-FALSE SELF

False Core Distractor Identies
I-dentity #1: There is no real wisdom
I-dentity #2: Over-idealistic or false optimism

I-dentity #1: No standards or rules
I-dentity #2: Super-standards and rules

Ichazo I-dentities
Ichazo I-dentity #1: Superior
Ichazo I-dentity #2: Inferior

FALSE CORE DRIVER—"I AM INCOMPLETE"

1. Notice where in your body you experience the *"I am incomplete" I-dentity*. Write down the answer until nothing pops up.

2. Now notice who, if anyone, modeled the *"I am incomplete" I-dentity*. Write down the answer until nothing pops up.

Therapeutic Note

At this point it is advantageous to externalize the *I*-dentity. This can be done in the following way:

Step I: Notice where in the present time body the *I*-dentity is located.
Step II: Notice the *I*-dentity's size and shape.
Step III: Take the label off of the *I*-dentity. Have it as energy.
Step IV: Allow the *I*-dentity as energy to move from the present time body to another physical location in the room, and then to become solid again.

3. Has this *"I am incomplete" I-dentity* ever been used in relationship to one or both of your parents? Ask the I-dentity this question and write down the answer until nothing pops up.

4. Has this *"I am incomplete" I-dentity* ever been used to compare itself with the same-sex parent and felt inferior, incomplete, or that there was something missing. For example, you felt like you couldn't compare to mom, or compare to dad. Ask the I-dentity this question and write down the answer until nothing pops up.

5. Has this I-dentity ever created innumerable plans or endless futurizing, when it continue to see or experience itself as either incomplete and inferior or complete and superior. Ask the I-dentity this question and write down the answer until nothing pops up.

6. Has this *"I am incomplete" I-dentity* ever been used in relationship to a guru or a teacher or projecting it onto a guru or teacher as they are incomplete? Ask the I-dentity this question and write down the answer until nothing pops up.

7. Ask the I-dentity to recall a time in its childhood when its parents kept on asking, "What are you going to do when you grow up?" and you felt like you could not make plans and, hence, felt inferior and to compensate you made over-planned to feel superior. Ask the I-dentity this question and write down the answer until nothing pops up.

8. Notice how young the *I*-dentity feels and how age-regressed it feels when it goes into feeling like it has to make plans in order to feel complete. Ask the *I*-dentity this question and write down the answer until nothing pops up.

9. Notice if the I-dentity has plans of being incomplete in the future or complete in the future. Ask the *I*-dentity this question and write down the answer until nothing pops up.

10. Notice how age regressed this *"I am incomplete"* I-dentity is and, hence, not grounded in present-time reality.

11. During the emergence→I-dentification with this *"I am incomplete" I-dentity*, was there any chaos or shock that occurred? Ask the I-dentity this question and write down the answer until nothing pops up.

12. Did the *I*-dentity ever feel incomplete at the physical, emotional, mental, spiritual, or sexual inferiority with its parents, siblings or your schoolmates in class? Ask the *I*-dentity this question and write down the answer until nothing pops up.

13. Has this *"I am incomplete" I-dentity* ever been unable to complete projects by having too many things started and not being able to reach completion on them? Ask the I-dentity this question and write down the answer until nothing pops up.

14. By taking on this *"I am incomplete" I-dentity*, what assumptions, beliefs, or decisions arose? Ask the I-dentity this question and write down the answer until nothing pops up.

15. Did this *"I am incomplete" I-dentity* ever have difficulty fitting into its family position, i.e., less than dad or mom or other brothers or sisters. Ask the I-dentity this question and write down the answer until nothing pops up.

16. Notice again where in your body you experience this inferior/incomplete I-dentity, and the empty label.

17. Peel the *"I am incomplete"* and empty label back.

18. Ask the *"I am incomplete" I-dentity* and the empty label, "What are you seeking more than anything else in the world?" and allow them to answer. Write down the answer until nothing pops up.

Chapter XII

Therapeutic Note

Notice how many psycho-spiritual schools focus on being complete or finishing

19. Notice the **SPACIOUSNESS** underneath the I-dentities and the empty label and experience the feeling of *pure* wisdom inside the **SPACIOUSNESS** of your **ESSENCE**. As you experience that sense of *pure* wisdom, notice the size and shape of the I-dentity and the label. Notice the **SPACIOUSNESS** that surrounds it and see the spacious as being made of the same substance as the I-dentity and the empty label while you continue to experience *pure* wisdom with no subject-object.

FALSE SELF COMPENSATOR—
"I HAVE TO BE COMPLETE"

1. Where in *your body* do you experience the *"I have to be complete" I-dentity*? Ask the I-dentity this question and write down the answer until nothing pops up.

2. Allow yourself to experience the *"I have to be complete" I-dentity*. Write down what happened until nothing pops up.

3. Now experience the *"I am in complete" I-dentity*, in your body and observe, and then slowly move from the *"I am incomplete" I-dentity*, in your body to the *"I have to be complete" I-dentity*, noticing what has to happen in your body. Write it down the answer until nothing pops up.

Therapeutic Note

At this point it is advantageous to externalize the *I*-dentity. This can be done in the following way:

   Step I: Notice where in the present time body the *I*-dentity is located.
   Step II: Notice the *I*-dentity's size and shape.

Step III: Take the label off of the *I*-dentity. Have it as energy.

Step IV: Allow the *I*-dentity as energy to move from the present time body to another physical location in the room, and then to become solid again.

4. Has this *"I have to be complete"* I-dentity ever been used to create a False Self of how great you were? Notice how the *"I am complete"* is acting "as if" I or you were or are complete. Ask the I-dentity this question and write down the answer until nothing pops up.

5. Has this *"I have to be complete"* I-dentity ever told itself stories about how great it was? Ask the I-dentity this question and write down the answer until nothing pops up.

6. Has this *"I have to be complete"* I-dentity, ever created fantasies about how great it was? Ask the I-dentity this question and write down the answer until nothing pops up.

7. Notice when this *"I have to be complete"* I-dentity gets into future fantasies and plans of how great it is or will be when or after it completes or experiences (*fill in the blank*). What happens to the physical body? Do you lose awareness of your body? Ask the *I*-dentity this question and write down the answer until nothing pops up.

8. Has this *"I have to be complete"* I-dentity ever been used to justify fraudulent behavior? For example, "There is no wisdom in the world anyway, what does it matter," or acting "as if" it were already complete? Ask the *I*-dentity this question and write down the answer until nothing pops up.

9. Has this *"I have to be complete"* I-dentity, ever been used to make it appear that your status was higher than another's, higher than it actually was, or that you knew more than you actually

Chapter XII

did? Or experienced more than you actually did? Ask the I-dentity this question and write down the answer until nothing pops up.

10. Has this *I*-dentity ever been attracted to a psychological system, a spiritual system or a Guru who claimed that this does it all, gives guarantees, claims superiority to other systems or creates hierarchies placing themselves on top of the hierachy with the highest status? Ask the *I*-dentity this question and write down the answer until nothing pops up.

11. Has this *"I have to be complete" I-dentity* ever been used to promote itself as sexually superior or that you can "complete" others either sexually, psychologically, biologically, thinking, emotionally, spiritually or seen others who do that? Ask the *I*-dentity this question and write down the answer until nothing pops up.

12. Has this I-dentity ever created a feeling of spiritual superiority? Like"I am more spiritual than you are." Ask the *I*-dentity this question and write down the answer until nothing pops up.

13. Has this *"I have to be complete" I-dentity* ever taken on status symbols, such as a Lexus or a better car or nicer clothes or flashy jewelry? Ask the *I*-dentity this question and write down the answer until nothing pops up.

14. Has this *"I have to be complete" I-dentity* ever been used as a way of resisting feeling something? Ask the *I*-dentity this question and write down the answer until nothing pops up.

15. Has this *"I have to be complete" I-dentity* ever been used as a way of resisting your real status or experience? Ask the *I*-dentity this question and write down the answer until nothing pops up.

16. Has this I-dentity ever listened to people's experiences and always one-upped them; that its experiences were better than

theirs, more than theirs, greater than theirs, or its understanding was more, better, or greater than theirs? Ask the *I*-dentity this question and write down the answer until nothing pops up.

17. Has this I-dentity ever joined a spiritual group or organization where hierarchy was the setup? (e.g., We have the highest, best or fastest way to God or enlightenment.) Ask the *I*-dentity this question and write down the answer until nothing pops up.

18. Has the *"I have to be complete" I-dentity* ever acted "as if" it were complete (finished) as the solution to a feeling of incomplete or there is something missing? Ask the *I*-dentity this question and write down the answer until nothing pops up.

19. Has the *"I have to be complete" I-dentity* ever been used as a way of resisting its own personal limitations? Ask the *I*-dentity this question and write down the answer until nothing pops up.

20. Has this *"I have to be complete" I-dentity* ever been used as a way of feeling special, different, superior, inferior, or more spiritual than another? Ask the *I*-dentity this question and write down the answer until nothing pops up.

21. Has this I-dentity ever made everything (no matter what) great, okay, etc., using reframing as a defense? Write down the answer until nothing pops up.

Therapeutic Note

The "I am incomplete" "I am complete" I-dentity, or" has a major glitch in it. In this situation, the shock and its pain are fused with *anything* that could be "seen" as painful, uncomfortable, or in a word, unpleasant. For this reason, they hallucinate and create" positives" out of or from anything to avoid the shock. This False Core Driver and False Self Compensator always are seeking pleasures or creating pleasures. In this way, reframing became an " ungrounded"defense

Chapter XII

and resistance against the realization of separation, and any painful life situations.

22. During the emergence→I-dentification with this *"I have to be complete" I-dentity*, was there any chaos or shock that occurred? Look specifically for the realization of separation and later being re-forced by sibling rivalries or rivalries with parents, or comparisons with siblings or parents. Ask the *I*-dentity this question and write down the answer until nothing pops up.

23. During the taking-on of this *"I have to be complete" I-dentity*, were there any decisions, assumptions, or beliefs that got created? Ask the I-dentity this question and write down the answer until nothing pops up.

24. Notice where the *"I have to be complete" I-dentity* is in your body.

25. Peel it back and notice how it covers the "I am incomplete" I-dentity, or *"There must be something missing" I-dentity* and the empty label.

26. Enter into the **SPACIOUSNESS** or **ESSENCE** and "look out" and view the False Core-False Self and the empty label.

27. As you look out at those two I-dentities and the empty label, notice what they are seeking. Write down the answers.

28. Experience the essential quality they both were seeking inside the **SPACIOUSNESS** of **ESSENCE**.

29. Notice the size and shape of the I-dentities and the empty label and **SPACIOUSNESS** that they are floating in. See them all as being made of the same substance, while you continue to experience the essential qualities of **ESSENCE**.

CONCLUSION

1. Without using your thoughts, memory, emotions, associations or perceptions, are you complete, incomplete or neither?
2. Without using your thoughts, memory, emotions, associations or perceptions, what does complete or incomplete even mean?
3. Without using your thoughts, memory, emotions, associations or perceptions, what does empty mean?

NOTICE THE NO-STATE STATE OF THE NON-VERBAL I AM.

CHAPTER XIII
THE WAY OF THE POWERLESS

FALSE CORE DRIVER:
I am powerless.

FALSE CORE COMPENSATOR:
I must prove I am not powerless by acting "as if"
I am overly powerful.

This False Core-False Self has the obsessive compulsive tendency to seek power and to resist powerlessness. This *might* be demonstrated in the *outward*[1] behavior of politicians and people whose general motivating principle *appears* from the "outside" to be "power."

For the infant this is, "I will enter Nirvana (merge with mom) if I am not powerless but powerful."

---

[1] It should be noted that it is very difficult to determine what is actually driving or motivating behavior (False Core Driver), it is, however, easy to determine the outward behavior (False Self Compensator).

The obsessive-compulsive tendency to type another is part of one's False Core-False Self Compensator. This behavioral age-regressed tendency is driven by the False Core Driver as an attempt by the False Self to gain control over a perceived chaos, which lies in the past only. This "mind-reading" and hyper-vigilant attempt is driven by the nervous system's *scanning-searching* device as part of its *survival* mechanism. Unfortunately, this tendency of the False Self Compensator is associated with this scanning-searching survival mechanism and is both aberrated and not in present-time.

Chapter XIII

REASONS FOR I-DENTITIES

I am powerless because I have no force, have no influence, got screwed over, etc. (*fill in the blank*).

DISTRACTORS

FALSE CORE: EMOTIONAL DISTRACTORS

Ichazo calls their underlying state a "venge," short for revenge, because people with this fixation are into revenge and rage. Quantum Psycnology sees this distractor as a past time reaction to the shock of the Realization of Separation from which the conclusion was drawn that "I am powerless." They are going to get even for the separation. For people into character analysis, this "I am powerless-I have to be powerful" often has a psychopathic tendency and a corresponding body type. This is because these people have such an unacknowledged feeling of powerlessness, they compensate by acting *overly* blown-up and imagining themselves to be much more powerful than they really are. Another distraction from powerlessness is turning love into lust. Love is warm and vulnerable. But for this False Core vunerability reactivates the shock of the Realization of Separation and they must resist this. Consequently, they move from experiencing vulnerable love to having the imagined "power" of lust.

This False Core Driver believes there is no real truth and overcompensates for this by becoming, according to Ichazo, over justice makers.

Ichazo I-dentities
I-dentity #1—hedonist
I-dentity #2—puritan

When you sit with a person who has this False Core Driver-False Self Compensator of overly powerful, you feel that there is no room for you. In other words, they demand that everything be organized around them. This deep feeling of powerless is seen in gurus or therapists who react in rage when a disciple or student or client feels powerful and decides to separate from them, to leave and be free. They react this way because if the disciple shows power, the gurus and therapists fall into the other side of the polarity and feel powerless thus reactivating their (the teacher, guru, or therapists') own shock of the Realization of Separation which triggers their rage.

This rage subsequently distracts them from the False Core Driver of "I am powerless" and so they have to act overly powerful. Unfortunately, gurus of this ilk do not see service to a teacher as a phase and a model which might lead to freedom. Quite the contrary. When students leave with a feeling of completion and empowerment, these gurus and teachers re-experience their own shock of the Realization of Separation. Therefore, they cannot allow separation and freedom or anyone else to "have it" otherwise they feel less powerful, less merged and more separate.

It should be noted that this is a strange reaction from teachers, therapists or gurus. given that they are supposed to help to set you free. However, some gurus simply cannot tolerate this happening and attempts to bind their students to them since another person's power triggers their own sense of powerlessness.

I once met a (so-called) "Master" whom we will call Big Baba. In his workshops, disciples were taught by Big Baba how to be a "Master" and earn money from others by turning them into "Masters." *Multi-level enlightenment*. When you completed his training, you had to sign a legal document which stated:

1. You would keep sending money to Big Baba.
2. Under penalty of law you could not share his secrets of "Freedom."

QUANTUM PSYCHOLOGY PRINICIPAL:

If psycho-spiritual systems are supposed to make you free, teachers should celebrate your freedom, not age regress and unknowingly compete with Dad/Mom or siblings for who is more enlightened (merged).

Chapter XIII

SPIRITUAL CONFUSION

"**I AM THAT**" at one level means **I AM THAT ONE SUBSTANCE**, and so are you (see Volume III). It does not mean "**I AM THAT** and you are not, but if you worship me, serve me, pay me, give me more, you too can become **THAT**. Thus, they have the power, and you have to play their game in order to merge with Mom/God. But really playing their game and not questioning are a re-enactment of the fusion response to Mom or Dad after the shock of the Realization of Separation.

> **Spiritual-Psychological Defenses:**
> Accumulating power, becoming the all-powerful God/Mom/Guru, accumulating "personal"power, getting what you want—are all attempts to fuse with magical mommy's *imagined* magical powers (omniscience, omnipotence, omnipresence, etc.).

FALSE POWER

Real power comes from within and it is not personal (see Volume III). There is no question that Gurus who know **THAT** might have *powers*. However, a true Guru knows that they as an individual are *not the personal source of that power*, nor do they identify with or see themselves that way. They are amazed, astonished, and surprised by what occurs or can occur around them.

There is no question that "miracles" occur around such people. However, they should realize that it is not personal to them. But because of "uncooked seeds," they get hooked into the adulation they receive and soon try to make it happen. For this reason, gurus will sometimes play tricks to keep this illusion of "I am powerful." One guru in India used to have his staff find out things about well known or rich people who were coming to meet him. Hours later when the well-known people arrived, he would say things (which his staff told him) which made him appear as though he were all powerful and magical (magical mommy) thus attracting even more disciples. In this case, the authentic power from within was substituted for a false act, a kind of spiritual manipulation which resulted in a collapsing of the levels. It is trying to create the illusion of spiritual power (which is not personal) by using external world manipulations to handle thinking, emotional and biological age regressions and the shock of the Realization of Separation.

THE PROTOCOLS: DISMANTLING THE FALSE CORE-FALSE SELF

Quantum Psychology's False Core Driver:
I am Powerless.

Quantum Psychology's False Self Compensator:
I must prove I am not powerless and that I am overly powerful.

Ichazo's I-dentity: #1—Hedonist
Ichazo's I-dentity: #2—Puritan

This next set questions are mini-meditations. The purpose is to ask yourself a question and notice what, if anything, pops up.

FALSE CORE DRIVER— THE I AM POWERLESS I-DENTITY

1. Who modeled this *"I am powerless" I-dentity* for you? Ask the I-dentity this question and write down the answer until nothing pops up.

2. Notice where in your body you experience the *"I am powerless" I-dentity*. Go into the "I am powerful" I-dentity and move gently from "I am powerful" to "I am powerless" and notice and study your body change in posture and movement. In this way you can get a sense for what it is like to move from the *"I am powerful" I-dentity* into the *"I am powerless" I-dentity*.

Chapter XIII

Therapeutic Note

At this point it is advantageous to externalize the *I*-dentity. This can be done in the following way:

Step I: Notice where in the present time body the *I*-dentity is located.
Step II: Notice the *I*-dentity's size and shape.
Step III: Take the label off of the *I*-dentity. Have it as energy.
Step IV: Allow the *I*-dentity as energy to move from the present time body to another physical location in the room, and then to become solid again.

3. Did this *I*-dentity ever punish another who it internally judged as bad for expressing powerful tendencies? Ask the *I*-dentity this question and write down the answer until nothing pops up.

4. Ask the I-dentity to recall a time its parents punished it for any kind of expression of power. Ask the *I*-dentity this question and write down the answer until nothing pops up.

5. Has this *"I am powerless" I-dentity* ever been used as a way to deny pleasure to itself to another? Ask the *I*-dentity this question and write down the answer until nothing pops up.

Therapeutic Note

Check to see if I cannot receive or give pleasure is fused with powerlessness.

6. What self-talk does this *"I am powerless" I-dentity* use. Notice if there is a fusion with a parent that says things like, you shouldn't feel *(fill in the blank)*. Ask the *I*-dentity this question and write down the answer until nothing pops up.

7. Has this *"I am powerless" I-dentity* ever been used to create a spiritual justification for denial of pleasurable sensations? Ask

the *I*-dentity this question and write down the answer until nothing pops up.

8. Has this *I*-dentity ever created a political philosophy to justify people being powerless? Or to criticize another for being into power? Ask the *I*-dentity this question and write down the answer until nothing pops up.

9. Notice any kind of church group or spiritual organization which advocates an *"I am powerless"* system or a model of denying sensation. (i.e., I am helpless (powerless) and I must depend on (*fill in the blank*)). Notice if you belong to any spiritual group that says things like, "Sex is bad, desires are bad, wants are bad, anything that inhibits the natural, energetic movement or the biological flows in the body is bad." If so, biological is fused with power and hence is bad, so that you must stay powerless. Ask the *I*-dentity this question and write down the answer until nothing pops up.

10. Has this I-dentity ever taken on a psycho-spiritual system which is into accumulating (outer power). Ask the *I*-dentity this question and write down the answer until nothing pops up.

11. Has this I-dentity ever taken on a spiritual philosophy that said that in order to experience the truth, you have to be powerlessness ("surrendered")? Ask the *I*-dentity this question and write down the answer until nothing pops up.

12. Ask the I-dentity to recall a time this *"I am powerless"* I-dentity felt proud of itself for not having a pleasurable time not yielding to desires or for denying itself sex. Ask the *I*-dentity this question and write down the answer until nothing pops up.

13. Has this *"I am powerless"* I-dentity ever helped to feel special, different, more spiritual from another (e.g., like you are extra good, extra pure)? Ask the *I*-dentity this question and write down the answer until nothing pops up.

Chapter XIII

14. During the emergence→I-dentification with this *"I am powerless" I-dentity*, was there any chaos or shock which occurred? Ask the *I*-dentity this question and write down the answer until nothing pops up.

15. During the emergence→I-dentification with this *"I am powerless" I-dentity*, what beliefs, assumptions or decisions were made? Ask the *I*-dentity this question and write down the answer until nothing pops up.

16. By taking on this *"I am powerless" I-dentity*, what emotions, sensations or shock gets resisted or is it resisting? Ask the *I*-dentity this question and write down the answer until nothing pops up.

17. Has this *"I am powerless" I-dentity* ever used energetic sensations, sought outward expression and allowed them to go up into the "head" yielding more thoughts or mystifying the world (i.e., like the world of gods and goddesses), or "you will go to heaven for denial of your bodily functions?" Ask the *I*-dentity this question and write down the answer until nothing pops up.

18. Notice where in your body you experience the *"I am powerless" I-dentity* and the empty label now. Write down the answer until nothing pops up.

19. Notice the size and shape of the *"I am powerless" I-dentity* and peel it back, and notice if the empty label is underneath it.

20. Notice if there might be an energy of rage that is in between the *"I am powerless"—"I have to be powerful" I-dentity*. Write down the answer until nothing pops up.

21. Now peel back the *"I am powerless" I-dentity* and the empty label, and go into the **SPACIOUSNESS** inside your physical body.

22. Ask the *"I am powerless" I-dentity* and the empty label, "What are you seeking more than anything else in the world?" (and allow them to answer). Write down the answer until nothing pops up.

23. Experience the essential quality of **ESSENCE** that the I-dentities and the empty label is seeking.

24. Notice the size and shape of the two I-dentities and the empty label and the **SPACIOUSNESS** they are floating in, and see them as made of the same substance while you experience the essential quality of **ESSENCE**.

FALSE SELF COMPENSATOR—
"I HAVE TO BE POWERFUL" I-DENTITY

1. Notice where in your physical body you feel this *"I have to be powerful" I-dentity*. Write down the answer until nothing pops up.

2. Notice what happens to your body if you go into this *"I have to be powerful" I-dentity*. Write down the answer until nothing pops up.

3. Who modeled this "I have to be powerful" pleasure-seeking *I-dentity*? Write down the answer until nothing pops up.

Therapeutic Note

At this point it is advantageous to externalize the *I*-dentity. This can be done in the following way:

Step I: Notice where in the present time body the *I*-dentity is located.
Step II: Notice the *I*-dentity's size and shape.
Step III: Take the label off of the *I*-dentity. Have it as energy.

Step IV: Allow the *I*-dentity as energy to move from the present time body to another physical location in the room, and then to become solid again.

4. Ask the *"I have to be powerful"* I-dentity when it felt incredible amounts of pleasure-seeking, and it was indulged immediately. For example, with lollipops or food or hugs or kisses. Ask the yourself this question and write down the answer until nothing pops up.

Therapeutic Note

Notice if you have fused together biological pleasure, i.e., sex, food, sleep, etc., with power or a separate self. If so, do the six step defusion process.

5. Has this *"I have to be powerful"* I-dentity ever been used to manipulate another? Ask the *I*-dentity this question and write down the answer until nothing pops up.

6. Has this *"I have to be powerful"* ever objectified another, making them into an object for pleasure? Ask the *I*-dentity this question and write down the answer until nothing pops up.

7. Has this *"I have to be powerful"* I-dentity ever been used to make others feel powerless (either inside yourself or through actions) so it could feel powerful? Ask the *I*-dentity this question and write down the answer until nothing pops up.

8. Has this *"I have to be powerful"* I-dentity ever gotten so absorbed in sensation that it forgot about the other person or the possible consequences to them? Ask the *I*-dentity this question and write down the answer until nothing pops up.

9. Has this *"I have to be powerful"* I-dentity ever created a philosophy to justify its actions and immediate gratification (i.e.,

like "going with it," or "It's just the body," or some kind of spiritual or psychological philosophy like tantra, etc.)? Ask the *I*-dentity this question and write down the answer until nothing pops up.

10. Has this *"I have to be powerful"* I-dentity ever been so fused with pleasurable sensation that you prematurely ejaculated or felt powerful because you could seduce a powerful person? Ask the *I*-dentity this question and write down the answer until nothing pops up.

11. Has this *"I have to be powerful"* I-dentity ever been used to alleviate frustration? Ask the *I*-dentity this question and write down the answer until nothing pops up.

12. Ask the I-dentity to recall a time when it got enraged and wanted or took revenge on a person who wouldn't fulfill your desires immediately. Ask the *I*-dentity this question and write down the answer until nothing pops up.

13. Ask the *"I have to be powerful"* I-dentity to recall a time it manipulated and power-tripped another to get its sexual or pleasure needs met. Ask the *I*-dentity this question and write down the answer until nothing pops up.

14. Has this *"I have to be powerful"* I-dentity ever been used to seduce another to get its needs met. Ask the *I*-dentity this question and write down the answer until nothing pops up.

15. Has this *"I have to be powerful"* I-dentity ever told itself that it loved somebody and changed love into lust so it could feel the power of lust rather than the vunerability of love. Ask the *I*-dentity this question and write down the answer until nothing pops up.

16. Has this *"I have to be powerful" I-dentity* ever felt special or different from another? Ask the *I*-dentity this question and write down the answer until nothing pops up.

17. During the process of identification with the *"I have to be powerful" I-dentity*, was there any chaos or shock that it was trying to overcome? Ask the *I*-dentity this question and write down the answer until nothing pops up.

18. During the process of taking on this *I*-dentity, were there any beliefs, assumptions, or decisions that were made? Ask the *I*-dentity this question and write down the answer until nothing pops up.

19. During the taking-on of this *"I have to be powerful" I-dentity*, were there any emotional states or feeling states resisted (i.e., feeling powerless, vunerable)? Ask the *I*-dentity this question and write down the answer until nothing pops up.

20. Notice again where in your body you experience this *"I have to be powerful" I-dentity*. Write down the answer until nothing pops up.

21. Peel it back and go into the **SPACIOUSNESS** of **ESSENCE**.

22. Ask the *I*-dentity and the empty label placed on **ESSENCE**, "What are you seeking more than anything else in the world?"

23. Feel the essential quality this *I*-dentity and the empty label are seeking.

24. Notice the size and shape of the *"I have to be powerful" I-dentity* and the empty label. Notice the **SPACIOUSNESS** that they are floating in.

25. See the *I*-dentity, the empty label and the **SPACIOUSNESS** as made of the same substance while you continue to feel the essential quality of **ESSENCE**.

CONCLUSION
1. Without using your thoughts, memory, emotion, associations or perceptions, are you powerless, powerful or neither?
2. Without using your thoughts, memory, emotion, associations or perceptions, what does powerful or powerless even mean?
3. Without using your thoughts, memory, emotion, associations or perceptions, what does empty mean?

**NOTICE THE NO-STATE
STATE OF THE NON-VERBAL I AM**

CHAPTER XIV
THE WAY OF THE LOVELESS

FALSE CORE DRIVER:
I am loveless; There is No Love.

FALSE SELF COMPENSATOR:
I must prove I am not loveless.
I must be lovable and loving.

The False Core Driver" I am loveless" or "There is no love" compensates with a persona or façade of a False Self which appears loving and accepting of what's happening. However, underneath this loving, accepting mask lies a passive, sometimes aggressive, coat of armor which is difficult to penetrate because of the on-going *spiritualization* and denial of "I am loveless, there is no love."

This False Self Compensator acts overly loving and seeks love (spirituality). As with all False Self Compensators, unfortunately, they can never make up in present time for the no-love which they did not get or decided was the reason for the shock of the Realization of Separation or what it meant. The shock of the Realization of Separation must be processed without intention and with awareness otherwise the overly loving act is always a defense.

Acting or appearing loving and lovable is their way to Nirvana (merging with mom). But as with all False Cores, they never get what they really want and always wind up subjectively with the experience of no love. This leads the "loveless" to a passive-agressive repressed anger. Why? Because the anger around separation is now generalized to the world and it can never come out directly because of the feared withdrawal of love from others as a consequence. In other words, merging with Mom=love—Separation=No Love. To express anger directly at No love would be" seen" as an unlovable act, one that would trigger the memory of the dreaded "shock of separation."

REASONS AND JUSTIFICATIONS FOR I-DENTITIES

I am loveless because I am (was) unlovable, I was (am) bad, empty or (*fill in the blank*).

This False Core is further compensated by the False Self in and through what Ichazo calls over-seeking. They have a tendency to be *over-spiritualized* types, looking for love in all the wrong places.

DISTRACTORS

This False Core has a core of underlying anger, which manifests itself with two I-dentities, one passive, the other aggressive. Passive-aggressive is a dual I-dentity, with a passive outside and an aggressive inside. The *I*-dentity acts "passive" and "projects" the aggressive onto another, and gets the other to "act out" the aggressive by frustrating it. In other words, it appears "as if" it is passive, but it really is trying to get you to "act out" in an aggressive manner.

At the same time, this False Self Compensator has a tendency to be overly spiritual (actually it is an act and a *spiritualization*) as a way to a-void lovelessness. In Quantum Psychology, the deep trance of *spiritualization*[1] is defined as taking an event, situation, emotion, thought or experience, and attributing it to spirituality or to some higher power like God, to a-void one's pain. They believe if they serve God (Mom), He will give them the love (i.e., the merger) that disappeared in the shock of the Realization of Separation. Go to a Buddhist center, yoga center or New Age bookstore and you will find many variations of this False Self.

---

[1] See *The Dark Side of the Inner Child*, Ch. 14, "Spiritualization."

Chapter XIV

THE PROTOCOLS: DISMANTLING THE FALSE CORE-FALSE SELF

False Core Distraction:
Inner States—anger, passivity.

Spiritual Defenses:
Poverty, loving, accepting, compliant (labeled as being surrendered).

Ichazo I-dentities:
I-dentity #1—Skeptic.
I-dentity #2—True believer, a devotee.

Quantum Psychology:
False Core Driver: I am loveless— there is no love.
False Self Compensator: I have to be lovable—acting overly-loving.

False Core Distractor:
I-dentity #1 - Passive
I-dentity #2 - Agressive

THE ENQUIRY PROTOCOLS

False Core Driver—The *I am loveless* I-dentity

1. Where do you feel this *"I am loveless, there is no love"* I-dentity in its body? Go into the posture of the I am loveless.

2. Who modeled this *"I am loveless, there is no love"* I-dentity for you? Write down the answer until nothing pops up.

Therapeutic Note

At this point it is advantageous to externalize the *I*-dentity. This can be done in the following way:

Step I: Notice where in the present time body the *I*-dentity is located.

Step II: Notice the *I*-dentity's size and shape.

Step III: Take the label off of the *I*-dentity. Have it as energy.

Step IV: Allow the *I*-dentity as energy to move from the present time body to another physical location in the room, and then to become solid again.

3. Has this *"I am loveless, there is no love" I-dentity* ever been used as a way of justifying why you feel separate from another and no-core? Ask the *I*-dentity this question and write down the answer until nothing pops up.

4. Has this *"I am loveless, there is no love" I-dentity* ever been used to express anger, but make it look like it wasn't angry, it was just (*fill in the blank*). For example, clarifying trying to get clear. Ask the *I*-dentity this question and write down the answer until nothing pops up.

5. Has this *"I am loveless, there is no love" I-dentity* ever been used as a way of justifying debunking somebody as if they have no real love or heart? Ask the *I*-dentity this question and write down the answer until nothing pops up.

6. Has this *"I am loveless, there is no love" I-dentity* ever been used to put itself outside of an ideology or system to re-enforce its separate no-love strategy? Ask the *I*-dentity this question and write down the answer until nothing pops up.

7. Has this *"I am loveless, there is no love" I-dentity* ever been used as a way of creating reasons why it cannot surrender or merge with another? Ask the *I*-dentity this question and write down the answer until nothing pops up.

Chapter XIV

8. Has this *"I am loveless, there is no love" I-dentity* ever been used to create an "us" and "them," or to create "Them" as the enemy? Ask the *I*-dentity this question and write down the answer until nothing pops up.

9. Has this *"I am loveless, there is no love" I-dentity* and then felt alienated or misunderstood? Ask the I-dentity this question and write down the answer until nothing pops up.

10. Has this *"I am loveless, there is no love" I-dentity* ever created the feeling of catastrophe? Ask the *I*-dentity this question and write down the answer until nothing pops up.

11. Has this *"I am loveless, there is no love" I-dentity* ever been used as a way of creating polarized thinking, i.e., all good (love, God, Guru) or all bad (no love, no Guru, no God or even the devil)? Ask the *I*-dentity this question and write down the answer until nothing pops up.

12. Has this *"I am loveless, there is no love" I-dentity* ever been used as a way of justifying anger? Ask the *I*-dentity this question and write down the answer until nothing pops up.

13. Has this *"I am loveless, there is no love" I-dentity* ever used sarcasm (anger) so you could knock somebody off their pedestal? Ask the *I*-dentity this question and write down the answer until nothing pops up.

14. Has this I-dentity ever sought love and then taken a vow to never fall in love again? Ask the *I*-dentity this question and write down the answer until nothing pops up.

15. During the process of emergence→I-dentification of this *"I am loveless, there is no love" I-dentity* that is cynical, was there any chaos or shock that occurred? Ask the *I*-dentity this question and write down the answer until nothing pops up.

16. During the process of taking on this *"I am loveless, there is no love" I-dentity*, was there any decisions, assumptions or beliefs that arose and got identified with? Ask the *I*-dentity this question and write down the answer until nothing pops up.

17. Has this *"I am loveless, there is no love" I-dentity* ever been used as a way of resisting the fact that you feel like there is really nothing or no one inside of you, which is why there can never be love? Ask the *I*-dentity this question and write down the answer until nothing pops up.

18. Notice now where the *"I am loveless, there is no love" I-dentity* is in its physical body. Write down the answer until nothing pops up.

19. Peel back the *"I am loveless, there is no love" I-dentity* and the empty label placed on the **SPACIOUSNESS** of **ESSENCE**.

20. Go deeply into the **SPACIOUSNESS** of **ESSENCE**, peeling back any layers other than pure **SPACIOUSNESS** that you meet.

21. From back inside the **SPACIOUSNESS**, how do those I-dentities and the empty label seem to you now?

22. Ask the *"I am loveless" I-dentity* and the empty label, what are you seeking more than anything else in the world? Allow them to answer and write down the answers.

23. Experience the quality of **ESSENCE** that they are seeking.

24. Notice the I-dentities and the labels (layers) floating in the **SPACIOUSNESS** in the distance.

25. Notice that what the I-dentities and labels are seeking is really "back there" in the **SPACIOUSNESS** of **ESSENCE**.

Chapter XIV

26. Turn the I-dentities around so that they can look into the **SPACIOUSNESS** of **ESSENCE** and see what they were seeking was actually inside (i.e., the love), etc.

27. Take the label off of the I-dentities and the labels, and allow them to be reabsorbed into the **SPACIOUSNESS** as energy while you continue to feel the essential quality.

28. Whenever you are ready, open your eyes, keeping part of your attention on the **ESSENCE** with its essential quality.

FALSE SELF COMPENSATOR: "I HAVE TO BE LOVING" I-DENTITY

1. Notice where in its body in you experience the *"I have to be lovable" I-dentity*. Ask the I-dentity this question and write down the answer until nothing pops up.

2. Notice who modeled this *"I have to be lovable" I-dentity* for you. Ask the I-dentity this question and write down the answer until nothing pops up.

Therapeutic Note

At this point it is advantageous to externalize the *I*-dentity. This can be done in the following way:

Step I: Notice where in the present time body the *I*-dentity is located.

Step II: Notice the *I*-dentity's size and shape.

Step III: Take the label off of the *I*-dentity. Have it as energy.

Step IV: Allow the *I*-dentity as energy to move from the present time body to another physical location in the room, and then to become solid again.

The Way of the Human • The False Core and the False Self

3. Has this *"I have to be lovable" I-dentity* ever been used to fit into a spiritual group? Ask the *I*-dentity this question and write down the answer until nothing pops up.

4. Has this *"I have to be lovable" I-dentity* ever been used to create the belief system of "love is all that matters" as a way of resisting things like feeling bad about itself, or bad about a situation? Ask the *I*-dentity this question and write down the answer until nothing pops up.

5. Has this *"I have to be lovable" I-dentity* ever been used to mystify or to make things mystical, like it was somehow the hand of God? Ask the *I*-dentity this question and write down the answer until nothing pops up.

6. Has this *"I have to be lovable" I-dentity* ever used reframing, i.e., learning lessons, Karma, etc., when things were painful? Ask the *I*-dentity this question and write down the answer until nothing pops up.

7. Has this *"I have to be lovable" I-dentity* ever felt age-regressed, i.e., younger than it was or is? Ask the *I*-dentity this question and write down the answer until nothing pops up.

8. Has this *"I have to be lovable" I-dentity* ever been used as a way to merge with or resist being separate from another? Ask the *I*-dentity this question and write down the answer until nothing pops up.

9. Has this *"I have to be lovable" I-dentity* ever said that love is good; and hate, anger, fear, jealousy, sadness, etc., is bad. Ask the *I*-dentity this question and write down the answer until nothing pops up.

10. Has this *"I have to be lovable" I-dentity* ever been used to resist the feeling that there was nothing inside, i.e., I have no love

Chapter XIV

and no interior. Ask the *I*-dentity this question and write down the answer until nothing pops up.

11. Has this *"I have to be lovable" I-dentity* ever been used to imagine its experience was the same as another's experience? Ask the *I*-dentity this question and write down the answer until nothing pops up.

12. Has this *"I have to be lovable" I-dentity* ever been used as a way of feeling that another was very close to you, or inside you, or next to you, as a way of resisting its own separateness? Ask the *I*-dentity this question and write down the answer until nothing pops up.

13. Has this *"I have to be lovable" I-dentity* ever been used as a way of making you feel like it was special and chosen by God for some kind of a mission? Ask the *I*-dentity this question and write down the answer until nothing pops up.

14. Has this *"I have to be lovable" I-dentity* ever been used to create the idea that it is magically protected? Ask the *I*-dentity this question and write down the answer until nothing pops up.

15. Has this *"I have to be lovable" I-dentity* ever been used to feel that it was entitled to get something because it was more spiritual? Ask the *I*-dentity this question and write down the answer until nothing pops up.

16. Has this *"I have to be lovable" I-dentity* ever been used to resist being ordinary (like everyone else)? Ask the *I*-dentity this question and write down the answer until nothing pops up.

17. Has this "I am lovable" "I have to be lovable" I-dentity ever projected onto teachers or gurus, magical (i.e., magical mommy)? Ask the *I*-dentity this question and write down the answer until nothing pops up.

18. Has this *"I have to be lovable" I-dentity* ever been used as a way of reinforcing its co-/dependency (i.e., God or the universe will take care of me)? Ask the *I*-dentity this question and write down the answer until nothing pops up.

19. Has this *"I have to be lovable" I-dentity* ever been used as a way to expect somebody to take care of it? Ask the *I*-dentity this question and write down the answer until nothing pops up.

20. During the emergence→I-dentification with this *"I have to be lovable" I-dentity*, was there any shock or chaos that occurred? Ask the *I*-dentity this question and write down the answer until nothing pops up.

21. During the emergence→I-dentification with this *"I have to be lovable" I-dentity*, what decisions did you make? Ask the *I*-dentity this question and write down the answer until nothing pops up.

22. Notice where the *"I have to be lovable" I-dentity* and the emptiness label are in your body now. Write down the answer.

23. Peel back the "I have to be lovable" I-dentity and the empty label, and enter into the **SPACIOUSNESS** of **ESSENCE**.

24. Ask the "I have to be lovable" *I*-dentity and the empty label, "What are you seeking more than anything else in the world?" Allow them to answer.

25. Experience the essential quality the *I*-dentity and the label are seeking within **ESSENCE**.

26. From inside the **SPACIOUSNESS** of **ESSENCE**, "see" the I-dentity and the labels (layers) and the **SPACIOUSNESS** they are floating in as well being made of the same substance.

27. As they dissolve experience the Essential quality of **ESSENCE** that the *I*-dentity and the label(s) were seeking.

LOVE FUSIONS

Love is such a potent experience and so intrinsic to our being that a lack of it can cause a fusion between love and some kind of action. For example, love can be fused with money, and another hot button area (no pun intended) fused with sex.

THE LOVE PROCESS

Rarely is love for love's sake experienced. Love is generally fused or associated with action, feeling, thoughts, etc. Since your ideas about a thing is not the thing, all your ideas about love are not love and must be acknowledged and discarded. In this way, you can have the Essential Quality of love which is love for love's sake, or love with no object.

EXERCISE

Pair up with someone. The purpose of this exercise is to notice and sort out your love fusions so you can have love for love's sake.

Part I: Facing a partner, Start off sentences with:
1. If I take in your love, I might feel *(fill in the blank)*. For example, if I take in your love, "*Then I will feel obligated.*" If I take in your love, "*Then I feel like I have to reciprocate.*" If I take in your love, "*Then I feel I'm being engulfed,*" or "*I have to withdraw or there'll be no me,*" etc.

Notice what pops up and discard it.

Part II: Pair up with someone. Start off a sentence with: If I feel love, then I have to *(fill in the blank)*.

For example, If I feel love then I have to "*Go along with what you say or do.*" If I feel love, Then it means "*I have to hug you.*" If I feel love, then it means "*I have to make dinner for you,*" etc. Then do the six-step fusion process.

Part III: Next run the six-step fusion process.

For example, if you fuse together love equals obligation for me, what are you creating, etc.? Then work the six-step fusion with your partner. We are trying to pull out the love fusions. What does love mean to you which, of course, is not it.

Love is a highly charged issue. It is important to love for love's sake, without it being connected to anything else, or with a "so that (*fill in the blank*)." Cutting the associational loop is imperative to stabilize this and all Essential qualities (see Volume III).

CONCLUSION

1. Without using your thoughts, emotions, memory, associations or perceptions, is there no-love, love or neither?
2. Without using your thoughts, emotions, memory, associations or perceptions are you lovable, unlovable or neither.
3. Without using your thoughts, emotions, memory, associations or perceptions, what is love?
4. Without using your thoughts, emotions, memory, associations or perceptions, what does empty mean?

**NOTICE THE NO-STATE
STATE OF THE NON-VERBAL I AM.**

CHAPTER XV
VOLUME II REVIEW

THE INTEGRATED FALSE CORE DRIVER:
The Chrystallized Ego.

False Core Drivers are theoretically driven by resistance to the shock of the Realization of Separation. They are protected from this realization by 1) a layer of amnesia or other trances;[1] 2) False Core Distractor I-dentities; 3) False Core emotional distractors; 4) the False Self Compensator; and 5) movements to other False Self Compensators or False Core Drivers to a-void the shock of the trauma.

This shock is so great that it is experienced by the nervous system as chaos. The False Core Driver is an attempt by the nervous system to make sense of that chaos after the experience has already occured and give it a conclusion, a story or reason for its having taken place. In this way, it gives reasons for the shock of the Realization of Separation and a solution so that it will not occur again (the False Self) if I prove (*fill in the blank*).

In this way the shock is experienced at a lower (older) brain level, the False Core-False Self at a newer ("cortical") level. The False Core conclusion and the False Self's solution to this false conclusion is held so strongly because 1) it represents the nervous system's survival generalizing process; 2) the False Self presents a (False) solution or path out; and 3) because I would rather feel my False Core Driver than the shock of the Realization of Separation.

---
[1] See *Trances People Live: Healing Approaches in Quantum Psychology*, and *The Tao of Chaos: Trances and Traumas*.

For these reasons, the False Self Compensator—which is the persona, act, mask, role, facade, etc.—becomes so integrated and appears to be so subjectively a part of you that you and others imagine it *is* you. It is the major integrated False Self, and like Narcissus who fell in love with his image—you fall in love with your False Self Compensator (the way you want others and you to see you). For this reason, you will fight, defend, act and—implicitly or explicitly—try to prove it *is* you.

This is where Quantum Psychology disagrees with most "psycho-spiritual" systems which promote:

1. Appreciating another's fixation and compensation because it subtly implies that people are their False Core Driver-False Self Compensator.
2. Stating that love and acceptance (virtues or healthy) are better than hate and anger (vices or unhealthy). This attitude and lack of understanding of **THAT ONE SUBSTANCE** can easily draw one into the False Self and its own re-enforcement of the False Core. *It is the False Self which unknowingly seduces one into the whole concept of* healing, transforming, reframing, reassociating, spiritual(izing), etc.

QUANTUM PSYCHOLOGY PRINCIPLE:

All False Selfs are "as if" acts. They are not real. They are done to protect and defend, and are a False solution to the False conclusion of the False Core Driver and the shock of the Realization of Separation.

For this reason the False Core Driver-False Self Compensator must be smashed by any means necessary or liberation from one's psychology is impossible. Nisargadetta Maharaj was a master of smashing the crystalized ego of the False Core Driver-False Self compensator.

When I was in India, I'd often see new people who came to the ashram wearing their "spiritual act." Their hands were in Namaste (prayer position). Everything was love, bliss, God and enlightenment.

Then, within a very short period of time, they'd come and talk to me. Something had happened. Their facades had "cracked" in some way and it was as if their body (particularly their face) was made of glass and that had shattered their view of themselves. When this occurred, what emerged was anger, resentment, rage, fear, etc.—all the feelings that had been buried under the amnesic layer which separated the False Core and False Self from the shock of the Realization of Separation.

I remember a father and daughter who attended one of my workshops. The man was very sweet and vulnerable. The woman told me privately that years before he had been an alcoholic and was very abusive to everyone. He had almost died and gone into alcohol treatment and over the years his personality had been transformed. Quantum Psychology sees this situation as, this man's ego was so crystallized that the humiliation of facing his alcoholism and treatment, and almost dying, smashed the crystallized False Self-False Core. This led him into the open-hearted vulnerability of **ESSENCE** with the Essential qualities of love and compassion.

Fortunately or not, because of external circumstances, most people never get their crystallized ego smashed and Essential qualities rarely become stabilized in one's awareness. People on spiritual or psychological paths often imagine it is about feeling bliss or love. But these Essential qualities cannot stabilize until the False Core-False Self goes through the process of dying and being smashed. This is why, within too many psycho-spiritual systems, we see so much politics, backbiting, and emotional repression. Because when the False Self begins to thin-out, the pain underlying the False Core-False Self remains, until it is directly confronted.

This False Core Driver-False Self Compensator is so powerfully integrated that we rarely question it; indeed, we're incapable of seeing or questioning it. Unfortunately, in our present time feel-good world, this type of confrontation is viewed as abuse rather than being held in the light of someone holding a mirror to your face so that you can see what you have been hiding and defending against.

When I first met Nisargadetta Maharaj, he smashed my "spiritual act"—which I did not even know I had. Nor did I know it was false because it was supported by the spiritual community. But when my "spiritual" False Core-False Self was confronted, I realized I had been in a *group trance* which did not serve "me" if "I" wanted to find out who "I" really was.

Frequently, gurus and teachers, because of their own crystallized I-dentity, are unable to "crack" their student's shell. If they do, the students might leave them like momma left them. These guru wannabes do a dis-service to their students by allowing them to "act out" the False Core Driver-False Self Compensator and then having them serve the guru's False Core-False Self. This demonstrates the teachers' and students' age-regression and trance-ference issues which justify their behavior with *spiritualized* philosophies.

THE INTEGRATED AND CRYSTALLIZED PERSONA

False Core Driver	**False Self Compensator**
1. "There must be something wrong with me	Acting "as if" they are perfect, and everything is perfect.
2. I am unworthy	Acting "as if" they have extraordinary value.
3. Inability to do	Acting "as if" they can do anything and every action is significant.
4. Inadequate-"unappreciated"	Acting "as if" they have it all figured out.
5. I don't exist - " unseen"	Acting "as if" they know, and their thoughts are significant.

Chapter XV

6. I am alone | Acting "as if" they have a connection to everyone.

7. I am incomplete | Acting "as if" they are happy and enjoy having experiences.

8. I am powerless | Acting "as if" they have incredible power or a way to get power.

9. I am loveless | Acting "as if" they are lovable and loving.

10. I am crazy | Acting "as if" they are clear, healthy, sane and appropriate.

11. I am unsafe | Acting "as if" they are safe and can make others safe.

12. I am out-of-control | Acting "as if" they are in control.

Of course, these statements are over simplified. However, notice what goes on your "internal life," what words continually come up. Try to find out what is most important to you and you will uncover your False Core Driver-False Self Compensator and the degree of its crystallization. For example, is the word "perfection" always in your thoughts, is the word "unseen" in your thoughts? Remember that this doesn't mean that if you use these words, they are your False Core-False Self. It does mean that you might have another False Self Compensator to defend against your False Core Driver than the traditional one assigned to it. If you do—trace it back.

379

EXERCISE
ISOLATING YOUR STYLE: WITH A PARTNER

Determine who is **PERSON A** and **PERSON B** (and then reverse).

PERSON A:	I Resist my False Core by (*fill in the blank*) to **PERSON B**.
PERSON B:	Tell me what you say to yourself to validate your False Core.
PERSON A:	Responds
PERSON B:	Tell me an emotion you use to distract yourself from your False Core.
PERSON A:	Responds
PERSON B:	Tell me a biological thing you use to distract yourself from your False Core.
PERSON A:	Responds.

In the above, one begins by "staying aware" of reactions, thoughts, etc. and how they lead to your False Core. Once that's seen, along with all of the other assumptions, then it can be discarded, leaving the *VERBAL I AM*, and then the **NON-VERBAL I AM** to naturally emerge.

TRACING YOUR FALSE CORE REVISITED

Now that we have gone through the False Core Driver-False Self Compensator in more detail, let us review:

When working with your False Core, try simply to be present, without repression or expression. Then, try to notice what is just below any impulse. Once you move from impulse to story or action, you are in your False Self. *If you can stop and trace back the impulse, action or story, etc., and stay in the VERBAL I AM just prior to the impulse you are into the **NON-VERBAL I AM**.* Here can can begin to cut the associational chain. Remember it's important to learn how to stop when, for example, someone calls you up and you're incredibly angry and ready to yell, dissociate, accommodate, etc. The

Chapter XV

idea is for you to *stop* just before that, to just be with it, and trace the reaction back to the False Core.

To paraphrase the Spanda Karikas,

> "If, at the moment of extreme anger, extreme sadness, extreme joy or if you are running for your life—if at that moment you can become introverted, you would experience spanda (divine pulsation).

You have to ask yourself, "What's the worst that can happen?" or "What is so bad about that?, What's underneath or behind this experience?" Often the answer to these questions has nothing to do with the external incident. Usually (unless there is no False Core Driver) it has to do with trying to find ways to distract yourself.

TRACKING THE FALSE CORE DRIVER THROUGH THE FALSE SELF COMPENSATOR

Let's say you have a False Core Driver of "There must be something wrong with me." To handle that you might become a perfectionist, or try to change the outside world. You can come up with many different variations and chains of association. However, if you can stop, and sit in the False Core Driver or just *prior* to it, the **VERBAL I AM or NON-VERBAL I AM**, you can cut the associational trance. In that cutting away, everything relaxes and the background of spaciousness moves into awareness and becomes foreground. Presently, the foreground is the False Core-False Self and the associational trance. As the amount of awareness that is placed on the False Core-False Self is decreased, they become background and the **I AM** and Beyond move forward.

QUANTUM PSYCHOLOGY PRINCIPLE:

> Experiences do not mean anything. Placing meaning on things or experiences is creating conclusions about what *is* which only takes you away from what is.

Trace back all your feelings, thoughts, memories and associations to the False Core Driver. Anything can lead you back because a person's entire personality is organized around it. The movements of the mind are: 1) created to resist the False Core; and 2) you take your False Core with you no matter where you are. If I have the False Core Driver of "worthless" to a-void the False Core Driver, I might move and take on another False Self Compensator on top of what I have as a way to prove my worth (the over-achiever, for example). I am going to become a *super achiever/performer* so that I can feel I have worth so that I can overcome and a-void the pain of worthlessness. But deep down inside I still feel *worthless*. I have taken the False Core of worthless with me. In any movement of the mind through associations, we do one of the six R's of the False Self—*resisting, resolving, reforming, re-creating, re-enforcing or re-enacting* our False Core because we are taking our False Core with you.

Although we filter everything through the False Core Driver and the False Self Compensator, it is difficult to determine what they are because of the pain and shock involved. In fact it is for this reason that we spend lots of time moving away from the False Core Driver to other False Self compensators.

But it is really important to know your False Core-False Self structure. Once you do, it doesn't matter where you move or what you do. All that matters is what *drives* the movement, (i.e., the False Core). In this way you can trace back through the trances to the False Core's driving structure. Knowing and dismantling your False Core simplifies the task of liberating your awareness because you will be dismantling the core of your psychology of who you think you are. Nisargadatta Maharaj emphasized two parts to the **I AM**—first, the **VERBAL I AM** *prior* to the False Core; and second, the **NON-VERBAL I AM** *prior* to the **VERBAL I AM**, i.e., without any thoughts, memories, emotion, associations or perceptions

QUANTUM PSYCHOLOGY PRINCIPLE:

Because you name something, you do not know it. You only know the name you gave it. This descriptive level of the ner-

Chapter XV

vous system confuses Korzybski's order of abstractions (i.e., the ideas about the thing, is not the thing you are referring to).

QUANTUM PSYCHOLOGY PRINICPLE:

Each description, experience, etc., is true only at the level it is being described.

In this way, the False Core description is true at the thinking level but not at what is actually happenning at the external, biological or Quantum level.

**THE DESCRIPTION IS NOT
THE THING IS IT REFERRING TO.**

Notice the False Core Driver that motivated the "mind" to move. Remember Freud's statement that "all trauma comes in chains of earlier similar events." No matter what it moves to, it is always moving to a-void the False Core.

If you have traumas that have not been looked at, then it will be more difficult to "get" stabilized in **ESSENCE, I AM**, etc. The energy from unprocessed material will pull your awareness "forward" and out of **ESSENCE** or **I AM**. The bottom line is, however, no matter what you are experiencing if you can trace and sit in the False Core and accept it without trying to get rid of it, it loses power.

THE MOVEMENTS:
A DISTRACTING DEFENSE

QUANTUM PSYCHOLOGY PRINCIPLE:

The movements of the mind are all re-enforcing or compensatory and are ways to re-enforce or resist the False Core.

We could spend forever talking about and observing the movements and compensations, etc. since, as Korsybski said, "You can always say more about what you said." But the question is, what is it that is driving the movement of "your" mind and what is pulling the chain of associations?

THE OBSERVER

There are multiple observers, however, observation is a first step. But observation does not yield liberation because the observer is part of the False Core-False Self complex. Therefore, as previously mentioned, it could be written "Observer-False Core-False Self." To go beyond the False Core-False Self, you must also go beyond the observer.

TRACING YOUR FALSE CORE EXAMPLES

Therapeutic Note

No matter what's going on with you (your story line, etc.), if you stay with whatever you are experiencing and trace it back, you will find the False Core. You might try to compensate for the False Core but it never works, it's like a groove on a record, it keeps playing the same song over and over.

Example #1—Tracing and Discovering Your False Core

Wolinsky: What is your experience?

Doris: What is my inner experience? It's disconnected.

Wolinsky: So when you feel disconnected, what's the worst of it?

Doris: The aloneness of it.

Wolinsky: When you feel that loneliness of it, is that the bottom line of it? Is that the worst of it?

Chapter XV

Doris:	It can feel like being out in space without an umbilical cord.
Wolinsky:	And what would be the worst of being out in space without an umbilical cord? What's the worst of that?
Doris:	*I'm forever nothing.*

Therapeutic Note
Just peel it back and ask, "What's the worst of it?"

Example #2—Tracing and Discovering Your False Core

Mike:	For me, it's like a pending death.
Wolinsky:	What's the worst of dying? What's the worst of that?
Mike:	Well, when I ask myself that, then I start to get more relaxed. Feel less distress.

Therapeutic Note
When you acknowledge your False Core, it is so familiar you can oftentimes relax into it. This is why meditation must be monitored by someone who understands that relaxation is not the purpose of meditation, although it may be a by-product. Relaxation can come with familiarity and re-enforcement of the False Core. That is why so many seekers are stuck. The purpose of meditation is to begin to open up awareness of the **VOID** or **I AM** or **ESSENCE**. If, however, meditation is being done from the False Self (which it inevitably has to) it reinforces the False Core.

QUANTUM PSYCHOLOGY PRINCIPLE:
When you are finally ready to meditate, meditation is no longer necessary.

Wolinsky:	So if you were to die, what would be the tragedy of that? What's the feeling like?
Mike:	My breathing changes. I feel a lot more relaxed.
Wolinsky:	So, if you were dead, what would that mean?
Mike:	That I would be alone and the worst of it is *I wouldn't exist.*

Therapeutic Note
The key thing again is to stay with that experience. Especially when it is extremely painful. Follow the pain until it starts to unravel. The issue is that it comes back.

Male voice:	My pain always comes back.
Wolinsky:	So, what is the worst of it coming back?
Mike:	The worst of it coming back is forgetting what to do.
Wolinsky:	And if you forgot what to do, what's the worst of that?
Mike:	Then I would be in the pain.
Wolinsky:	And what's the worst of that?
Male voice:	I would be inadequate (begins to cry).

CONCLUSION

Example #3—Tracing and Discovering your False Core

Client:	I feel powerless.
Wolinsky:	What's the worst of the powerlessness.

Chapter XV

Female voice:		Something will happen.
Wolinsky:		What's the worst of that?
Female voice:		I would be responsible.
Wolinsky:		Since you can't feel responsible, you said it was too much for you, what's the worst of it?
Female voice:		I would feel inadequate or worthless.
Wolinsky:		Why don't you sit in either one of those.

Therapeutic Note

The word is just the description of the feeling state, but it's not the feeling state itself. So do not get too caught up in the word. The most important thing here to get is what is the thing that is pulling the whole chain of events. What's pulling the whole chain is the core of it. If you get that, everything else will loosen but if you don't get that, everything else will continue. I'm calling it a False Core because if you get to your False Core and peel it back your real **ESSENTIAL CORE** emerges, (see Volume III).

Therapeutic Note

You can get to the False Core by following deep feelings. Then notice how pervasive it is. The False Core is so unquestioned. It is almost invisible.

EXERCISES
RESISTING MY FALSE CORE

These are incomplete sentences to be done in pairs or with oneself. Keep asking the questions, writing down the answers until nothing else emerges.

1. When I feel my False Core, I (*fill in the blank*).
2. To not feel the False Core, I (*fill in the blank*).
3. The way I avoid my False Core is (*fill in the blank*).
4. To compensate for my False Core, I (*fill in the blank*).
5. To get away from my False Core, I (*fill in the blank*).

6. To not feel my False Core, I tell myself the following lies (*fill in the blank*).
7. To not feel my False Core, I go into the future and imagine (*fill in the blank*).
8. To not feel my False Core, I go into the past and recall (*fill in the blank*).
9. To not feel my False Core, I forget (*fill in the blank*).
10. When I get into my False Core, I can't see (*fill in the blank*).

Example #4—Tracing and Discovering your False Core

Client: I'm wanting to really avoid this experience of like worthless, faceless, like being totally exposed and everybody seeing that I'm this absolute, big nothing. I'm not sure.

Wolinsky: What is so bad about having no value?

Client: Well who would want to be with me?

Wolinsky: What is so bad about no one wanting to be with you?

Client: To feel this worthless and unimportant and alone.

Wolinsky: What's so bad about feeling worthless and alone?

Client: I wouldn't exist.

Wolinsky: What's so bad about non-existence?

Client: I would be alone—that's it.

Therapeutic Note

The story of what happens describes the movement. That story takes you out of the initial False Core that's pulling the whole chain of associations. What's pulling the whole chain of association is, "I don't want to feel this (I am alone)." That's the False Core. Underneath the False Core, you will notice that there is the **BIG EMPTINESS**.

Example #5—Tracing and Discovering your False Core

Male Voice: I feel alone.

Wolinsky: What's so bad about feeling alone?

Male Voice: Partly I feel protected in that. So it is a safe place as well as a place of—I guess the worst of that is I will never get out of that. I will always be like that.

Wolinsky: So you will always be isolated and alone. So what's so bad about that?

Male Voice: I would not have any love.

Wolinsky: Is that it?

Male Voice Yes.

Therapeutic Note

Stay with the False Core and see where it leads you. The easiest way to find out what is the False Core is to notice what you do not want to experience. What is it? It is humiliation, being exposed as stupid, etc., you don't know what it is. But what is it, no matter what, you really don't want to experience it. You can trace things back and get a sense of it once you are there. Be with it, focus on it. If you stay in that thing that pulls your chain of associations, that feeling (sensation) state, all of the traumas, emotions, beliefs, and all of the structures will pop up out of it. Every thought. is going to be your pointer or arrow leading to the False Core.

STEPS TO GOING BEYOND THE FALSE CORE

Step I.   Own the False Core.
Step II.   To be able to be it and observe it. (Before I was looking at the world through just this False Core.)
Step III. Being able to put it down (UN-BE IT).

THE BIOLOGY AND THE FALSE CORE DRIVER

At a biological level when the False Core is threatened, the biology feels a contraction. That is its defensive protective reaction to the False Core Driver. The False Core is an attempt to hide the belief that there is nothing (small n) or *no core*. In this way the False Core's survival is central to the survival of the organism—or so the False Core *thinks* and imagines. The False Core Driver is fused with the biology and they work together. This is part of the defense of the False Core. The biology contracts around your False Core. The False Core Driver-False Self Compensator (as part of the nervous system) has a survival mechanism.[2] What makes thoughts, emotions, beliefs, etc., so persistent is that they are part of the nervous system and, hence, have a survival instinct and a fight/flight mechanism contained within them. Furthermore, this biological-psychological contraction occurs when there is a threat to revealing not only the shock and the narcissistic wound, but also the belief that there is *no core* or nothing underneath or behind it to enable the bio-psychology (an individual self) to survive. The biological contracting protects, defends, and keeps you away from *no-core* and the realization of separation with its wounding shock.

WORKING WITH THE FALSE CORE-FALSE SELF

Knowingly, consciously and intentionally create the False Core (lie), create the aloneness, create the unworthiness, etc., and then the over-compensation. In other words, if you are in worthlessness, with full awareness of the fact, knowingly, consciously and

---

[2]The survival mechanism of fight/flight also contains over-generalization and the searching-scanning device

Chapter XV

intentionally create worthlessness—and proving worth—uncreate it, create it, uncreate it, create it. Make it as big as the room, as big as the universe, etc., without the intention of getting rid of it.

CONCLUSION

I think that the most common question people ask is, "Will my False Core disappear forever? The answer is Yes! But under stress it will come back, although *you* won't care. It may or may not even be functioning, but you still won't care. Other people might care. Other people might be in reaction to it, you won't. When your I-dentification with your False Core-False Self dissolves, you can be who you are and stabilize in **ESSENCE**. In fact other people who are stuck in their False Core-False Self might see you in yours, even if you are not.

In Raja Yoga, they say, "When a pickpocket sees a saint—they only see his pockets."

In Quantum Psychology, when you are in your False Core-False Self, you only see others in their False Core-False Self—NOT THEM.

To explain further, let's say your False Core is "worthless." As your awareness becomes less fixated and more multi-dimensional, suddenly you become more aware of your body and of others (the external). Your awareness deepens and expands. In Quantum Psychology this, is called the "development of functional awareness." Now, at first your fixations are in the foreground and you identify with your False Core and the material that comes out of it. But there is a shift. As the False Core moves to the *background*, it loses its power and is just not important any longer. At the same time, as functional awareness develops, **ESSENCE** and **I AM** move *forward* from the *background* and reveal themselves.

Supposing a man walked by the window over there. If my False Core were in the foreground, I might be asking myself, "What does he want? What's he thinking? Is he going to bother me? But now, it doesn't matter anymore. I don't really think about it, I simply notice his coming and going. Well, this man is a symbol of the False Core-False Self which now comes and goes and no longer has any effect on me. Because I know he (any *I* thought) is not me. A disciple once asked Nisardadatta Maharaj, "Does anything ever come up for you anymore?" "Yes," he said, "But I immediately see it is not me and it disappears."

CHAPTER XVI
THE FALSE CORE AND ITS TRIGGERS

DIFFERENTIATING THE LEVELS

It is important to know what triggers the False Core since the trigger it is the external level which re-activates the past. This trigger can lead you beyond the False Self-False Core to discover and help to take apart the False Core-False Self. Therefore, we need to: 1) See the trigger within an event; 2) slow the process down; and 3) study it using your awareness. For example, if seeing your boss triggered your anxiety, you need to ask yourself what triggered it? Was it his look, his words, his body language? Be specific. Triggers can also be multi-dimensional. Let's say the boss looked at you. That's the external dimension. At a thinking level, you had a thought called "He does not like my work. At a thinking dimension, you ask yourself, "What are my thoughts about that? Do I have trance-ference? What does this event remind me of? What does it bring up?" He raised his voice slightly (external level). At the emotional dimension, what was my response? His face was flushed, he was scowling (external dimension). What was my biological response? Did I collapse? Tighten my jaw?

These are some of the questions you need to ask yourself. If you can learn to break it down into smaller pieces and slow down the process by being aware of the triggers, then you can learn to see how the different dimensions interact. This liberates you from being fixated and makes multi-dimensional awareness accessible, hence **ES-**

SENCE is revealed. Why? Because your awareness will not be gobbled up disproportionately in any one dimension.

In this way, you can distinguish multi-dimensional awareness and the impact of a multi-dimensional trigger.[1] For example, in response to an event, experienced on a thinking level, you might have thoughts of, "I don't really care"; but at an emotional level, you are really *pissed off*, and at a biological level, you might have a physical reaction.

EXERCISE

1. Review at night or make a diary of your "reactions." Trace them back through the False Self Compensator to the False Core Driver.
2. Notice the patterns.
3. See how often the same False Core Driver and False Self Compensator patterns arise and in which' contexts (external dimension).

THE EXERCISE

Step I: What triggers your False Core more than anything else in the world?
Step II: What happens at a thinking dimension?
Step III: What happens at the emotional dimension?
Step IV: What happens at the biological dimension?

QUANTUM PSYCHOLOGY PRINCIPLE:

When there is so much attention and awareness placed on one dimension, the other dimensions of your humanness are going to be neglected.

Many kinds of psychotherapy and "spiritual work stay focused in *only one or two dimensions*. Practitioners in these fields

---
[1] This was the primary understanding in the Post Traumatic Stress Disorder Process (see *Tao Of Chaos*).

Chapter XVI

tend to perceive things through their own particular lenses, seeing only their dimension as the cure all. The bodyworker sees the physical body, the psychoanalyst looks to the mind's associations, etc. Part of the problem is the prior teaching and misinformation they have received, based on the idea that their own particular area does it all. Another difficulty is their naivete and unfamiliarity with other systems at the same time that they ignore the limitations of their own system.

The habitual "this does it all" approach simply re-enforces the old cliche, "If your only tool is a hammer—then every problem is a nail." In other words, it doesn't work.

Instead, it is essential to trace back. Movements and associations are a defense and a distractor of the False Core driver. The False Self's distracting story is made up, and might even contain, interesting material, but stories only beget more stories. Stephen Jourdain once said to me, "These stories and ideas are all very interesting, but if they bite my leg, I have to kill them."

> Quantum Psychology would say, "These stories are all interesting; however, they come after the experience itself has already occurred and are made-up creations of the Nervous System. And if they shrink my awareness, I must dismantle them (kill them)."

To locate and trace the False Core, you need to ask, "**What was created in response to the external**?" For example, what did I do when someone said "*X*"? Did I go into withdrawal? Then you need to ask yourself, "What's the worst of it?" or "What is so bad about that?" "I don't want to feel inadequate," "I don't want to feel worthless," "I don't want to feel I don't exist," etc. Trace the reaction or False Self Compensator back to the False Core which is the driver of your psychology.

QUANTUM PSYCHOLOGY PRINCIPLE:

The degree to which you are *willing to sit in your False Core* will be the degree to which you are *able to sit in your* **ES-**

SENCE. If you cannot sit in your False Core, then you cannot stabilize in **ESSENCE**.

QUANTUM PSYCHOLOGY PRINCIPLE:
Quantum Psychology is not interested in manifestations (i.e., the stories and explanations of the False Self). Quantum Psychology is interested in what drives the stories the False Core Driver.

QUANTUM PSYCHOLOGY PRINCIPLE:
The more dimensions you are aware of, the greater the subjective experience of freedom.

The less the collapsing of the levels, the easier the False Core-False Self is to work with. Differentiating the trigger helps to shift the percentage of awareness used for each dimension as well as the triggered response of the False Core.

EXERCISE
1. When reviewing a triggered response within a situation write down what percentage of your awareness went into each dimension.

For example, if your False Core is "inadequate," what triggers that response? Is it someone saying, "You're stupid?" What happens then, and what percentage of your awareness goes to the thought level? What about the emotional feeling level or the biological level? What happens there?

DEMONSTRATION
Wolinsky: Tell me something that triggers your False Core. (I want to get what triggers the False Core, and

Chapter XVI

	I'm going to work on through the different dimensions.)
Student:	When I am in the False Core, what triggers the False Core is that I am being assessed and observed critically.
Wolinsky:	Look out here and imagine you being assessed. Who in your past has observed you critically?
Student:	My mother was very critical.

Therapeutic Note

Here is the external level and the trance-ference of mom onto the external world.

Wolinsky:	The judgmental comes at you, and what emotion gets created in response to that?
Student:	Anxiety, nervous, anxious but not in a concerted way.
Wolinsky:	With these thoughts coming at you, emotional level, you get anxious and at a biological level, what happens?
Student:	I perspire.
Wolinsky:	You are imagining that the observed critical thoughts are coming from the group, what, if anything, happens in your **ESSENCE**?
Student:	Working in front of a group makes me nervous. Anxious.
Wolinsky:	Now, look at the (imagined) judgmental coming at you, just through the feeling function.

Student: I can feel the anxiety through the feeling function.

Wolinsky: Now, look at the group through the animal level, the biological animal level.

Student: I feel like turning away, hiding, freezing up, going inside. It's hard to look from the animal level. I have a hard time doing it.

Wolinsky: So you dissociate from the animal level. Look at the group from the essential level.

Student: It's fine to do that. It's all right. We can do that.

Wolinsky: Now, what I would like for you to do is to split your attention. First, we are looking through lenses. So I'm going to give you a lens. This is just the thinking lens. You are looking through the thinking lens at these critical people. Now look at them just through the emotional lens. Now look at them through just the biological lens if you can.

Student: It's hard to do that.

Wolinsky: When you go into the biology, the animal part, do you dissociate?

Student: I can do that but my mother is there censoring it.

Wolinsky: Now, I am going to give you the lenses real quick. Do them in any particular order. Look at them (the group) through the emotional level. Let me know when you've got that. Good, now look at them through a thinking dimension. Now look at them through an **ESSENCE** dimension. Now animal dimension. Now I want you to do it very fast. So

Chapter XVI

	thinking, feeling, biological, essential, biological, thinking. Now increase the spectrum and look at them through the thinking, the feeling and the biological. See if you can get a sense of the awareness through all the dimensions.
Student:	It feels more like an essential feeling.
Wolinsky:	How are you feeling now?
Student:	Very good.
Wolinsky:	Now, let's have a trigger for the False Core when the group looks at you critically just for a second. As they do, just very quickly inside yourself, *scan*. So you want to go thinking, feeling, just to check the different dimensions for yourself noticing the reactions in each dimension. What, if anything, gets triggered? I want you to do all the dimensions if you can. How are you feeling now?
Student:	Yeah, I feel all right. I feel okay. That doesn't say what I'm trying to say.
Wolinsky:	So right now as you feel these guys look at you critically, do you think inadequate?
Student:	No.
Wolinsky:	Do you feel inadequate?
Student:	No.
Wolinsky:	Does your biology perspire and heart accelerate?
Student:	No.

Wolinsky: Do you have an essential sense that's not available to you?

Student: No.

Wolinsky: How are you doing now?

Student: I feel great.

Wolinsky: How does the False Core, which normally gets triggered, how does it feel to you now?

Student: It's not triggered.

Wolinsky: Do you have anything to say?

Student: No. I want to simply digest it.

Wolinsky: How does the experience seem to you in a thinking dimension?

Student: Well, that the path or the process was one of experiencing and distancing and going back and forth, and taking it through my mother's lens and to the point now that there is no charge on it.

Wolinsky: How does the experience seem if you look at it from the emotional lens?

Student: There were periods that I was very nervous, very anxious, very self conscious, exposed but now I don't experience that.

Wolinsky: And how do you experience it from the biological window?

Chapter XVI

Student: I was feeling, I mean I was perspiring. I had an accelerated heart beat *but not now.*

Wolinsky: And through the essential level?

Student: Well, when the other experiences were very intense, it was hard to access the essential level, but now it's not difficult, **ESSENCE** feels very available.

CONCLUSION

Remember we want to be aware of what present time situations trigger past time associations. You can have a feeling or think something without it *meaning* anything. You want to take meanings and associations off of things. What happens is what happens. Usually there is a story about what it means; but, in reality, *experiences mean what you decide they mean.*

During a trauma, all the levels collapse. That is why it is so difficult. All the levels must be differentiated. You can, however, allow clients or yourself to go into **ESSENCE** and experience **ESSENCE** and then experience their mother's **ESSENCE** so that they can have an **ESSENCE** to **ESSENCE** experience. Simultaneously, you want to have a biological as well as a psycho-emotional *separation* but still maintain the experience of **ESSENCE** to **ESSENCE** at a different level.

The process of differentiating the trigger's importance can be seen in the following way:

1) You are differentiating the trigger like you would in a Post Traumatic Stress process;[2] 2) you are developing functional awareness; and 3) you are differentiating the collapsed levels by viewing the world, people and/or things through any one or a number of dimensions.

This leads to a greater flexibility and a functional awareness, (i.e., your ability to move, see and experience through a different dimension) which enhances your subjective experience of freedom.

---

[2] For a detailed view of Post Traumatic Stress Disorder, see *The Tao of Chaos: Quantum Consicousness, Volume II*

CHAPTER XVII
FORCE THEORY

"For every action there is an
equal and/or opposite reaction."
Issac Newton

THE TAO OF CHAOS, REVISITED AND REVIEWED

Traumas

The False Core has associational chains which act like a filter through which all traumas are perceived, ordered and experienced. These traumas are then generalized to different externals and re-used to reinforce and justify the underlying structure.

The False Core is the lens of interpretation. If you were molested at the age of eight, and your False Core were "There must be something wrong with me," the interpretation of that experience and what it means to you will be used to interpret all experiences throughout your life. In traumas there is a collapsing of the levels. Prior to trauma, everything has motion—thoughts, memories, sensations, emotions, and body movements. In fact, to have any experience, neurons of the nervous system *must* move. When a trauma occurs, three things happen: 1) Motion ceases; 2) memory is frozen; and 3) the levels collapse. This goes on in order for the nervous system's survival mechanism to have a generalizing response of, "I will not let this occur again," and a scanning-searching device to make sure "it will never happen again," and if it appears to or might appear to, I will be ready, with fight or flight. As mentioned, many times the over-generating and scanning-searching are *NOT* in present time—hence, they are inaccurate.

Traumas are obsessively-compulsively gone back to and re-lived; thus, collapsing of the levels. This:

1. Freezes the dimensions causing a picture memory to be formed.
2. Causes an obsessive-compulsive tendency to go over it again, to relive its pleasure or pain, and right the wrong.
3. If the frozen memory is unpleasant, it re-enforces the False Core driver.
4. If the frozen memory is pleasant, it can re-enforce the pain of the False Core driver.

The P.S.T.D. process[1] takes apart the levels and moves us onto the next step which needs to be addressed—*Force Theory*.

REVIEWING PRINCIPLES

QUANTUM PSYCHOLOGY PRINCIPLE:
You can *never* overcome one dimension with another.

QUANTUM PSYCHOLOGY PRINCIPLE:
Meet the problem at the level of the problem.

QUANTUM PSYCHOLOGY PRINCIPLE:
In trauma, all levels collapse. This yields a distortion in both the external and internal dimensions

It is the interpretation and story lines created in response to the external which reinforces the False Core and leads to 1) your

---

[1] See, *The Tao of Chaos*.

Chapter XVII

perception of world; 2) your perception of self; and 3) how you imagine the world perceives you. As long as the False Core is operative, you will always reinforce it. You will always set yourself up to have that internal subjective experience.

FORCE THEORY:
THE FALSE CORE DRIVER AS A FORCE—
THE FALSE SELF COMPENSATOR AS
A COUNTER-FORCE—FORCE

In force theory, a trauma can be compared to a force coming at you. Somebody might be trying to do something to you in some way. The question after the levels are differentiated is, "What did you create in response to this force?" If you have a force coming at you, you'll then create a force to counter it, *thus fusing them together as a unit*. Next step is then, "What did you decide about that?"

THE FORCE THEORY PROCESS

Step I You have a trauma (a force coming at you in some way).
Step II You counter that force by creating a response to it.
Step III What did you decide about that?
Step IV What is the worst of it?
Step V Go into the verbal **I AM** prior to the False Core.
Step VI Go into the non-verbal **I AM** prior to the False Core driver.

If you do this, it will ultimately bring you back into the False Core Driver which pulls your chain of associations by asking, "What's the worst of it?" In any situation, this process is a means to find your way back to the False Core structure so that you can go beyond it.

Take this example—imagine you go to return a shirt and the sales clerk won't take it back and acts hostile. I can lose my temper or feel really disappointed. What did I decide about it? I decided I feel unloved. What's the worst of it? I feel powerless. What the worst of it is, "I feel unloved," etc. That's the False Core driver.

To Illustrate

Step I	The force: Sales clerk, "I won't take it back."
Step II	Counterforce: Anger, disappointment.
Step III	I decided: I feel unloved.
Step IV	The worst of it: The False Core driver "I am loveless"
Step V	Go into the verbal **I AM**—prior to I am powerless
Step VI	Go into the nonverbal **I AM** prior to the False Core driver.

QUANTUM PSYCHOLOGY PRINCIPLE:

The False Core Driver-False Self Compensator interprets what an experience means to you. Therefore, the feeling that you have about an event and the "you" in the event are your False Core Driver-False Self Compensator.

GROUP EXERCISE

Step I: **PERSON A** to **PERSON B**—"What do you want (force) more than anything else in the world?" Or you could say, "What is it that I really want from taking this workshop?" **PERSON B** responds:

Step II: **PERSON A** to **PERSON B**—"What is preventing you from having that (counter-force?)" For example, if I want the ultimate truth, What's preventing me from having that? **PERSON B** says, "I'm not worthy. I'm inadequate," etc. Notice if the *ultimate truth* (in this example) is a False Self trying to overcome the False Core driver? Notice that if "it" (the *ultimate truth*) happens, does it ignite (trigger) the False Core driver?

Chapter XVII

QUANTUM PSYCHOLOGY PRINCIPLE:
In order to find out what you want, you might first have to go through a lot of what you don't want.

Example

Client: If I were in **ESSENCE** now, it would be much more pleasant.

Wolinsky: So what you are seeking (Force) is to be in **ESSENCE**. What would that experience be like? (Force)

Client: It would be more spacious.

Wolinsky: What is preventing you from having that? (Counter-force)

Client: My disappearance into details. (Counter-force)

Wolinsky: The disappearing into details identity is preventing you? (Counter-force)

Client: So if what I wanted was peace and what keeps me from it is fear, those are two identities?

Wolinsky: Yes.

Client: Peace is an identity? (Force) Fear is a counter-force.

Wolinsky: Seeking peace is an Identity. And preventing peace is the second *I*-dentity. (Counter-force).

Client: Actually I am seeking wholeness and worth (Force).

Wolinsky:	And what prevents you from getting that. (Counter-force)
Client:	It is my sense of worthlessness. (Counter-force)
Wolinsky:	Okay, that's two identities. I'm trying to create wholeness and/or worth and what prevents me is fear these will give you two very powerful identities to look at.

(To another person)

Sue:	Seeking love, play, balance (Force), and what's preventing me is I don't know how (Counter-force).
Wolinsky:	So if you had love, work, and play, what would that experience be like?
Sue:	Heaven. Oh, it would be just wonderful.
Wolinsky:	What would that wonderful feel like or be experienced as?
Sue:	Lack of stress. It would be like enough. Happiness.
Wolinsky:	And what prevents you (Counter-force)?
Sue:	I don't know how (Counter-force).
Wolinsky:	Okay. So that's the two identities. "Seeking happiness" (Force) and "I don't know how" (Counter-force).

CHAPTER XVIII
THE FALSE CORE KITE

The False Core kite is a simple but clear illustration of the way we can move away from the False Core. Let us first review some details before we explore and go through this process:

1. The False Core driver is the engine which drives your psychology.
2. The False Self Compensator is your attempt to overcome the False Core driver. It is a persona, belief or philosophy which feels that if I (*fill in the blank*), then I will attain merger with mom. This merger can later be trance-ferred onto God, enlightenment, etc.
3. The False Self and the distractor I-dentities frequently become a major vehicle to attain the illusion of merger with mom/God.
4. A trance and/or an emotion (generally containing some amnesia) separates and creates the *split off* between the False Core Driver and the illusion of the desired destination of Nirvana (merger with mom) which the False Self Compensator strives for by trying to overcome the False Core Driver.
5. You can go to *any* other False Core Driver-False Self Compensator or distractor emotions because if your False Core is "I don't exist," for example, it is better to feel powerless than to feel non-existant. This is true for any I-dentities or any False Self Compensators. I can try the False Self Compensator to be *perfect* so as not to feel the False Core of incomplete as a way

to get to Nirvana (merger with Mom). However, this movement keeps it all going like a machine.
6. All movements of the mind are generated to resist the False Core.
7. Through all the movements, you always take your False Core with you. This reminds me of a woman who went to Baba Muktananda telling him she wanted to go on vacation. Muktananda replied, "Tell me where you can go and not take your mind with you and I will go there."

Hopefully the illustrations will clarify.

MOVING TO ANOTHER FALSE CORE DRIVER →FALSE CORE DISTRACTOR

A False Core Distraction is a False Core which someone moves to in order to avoid their *actual False Core*.

The False Core distractor provides a *good reason* why you are your False Core. The False Core of "I don't exist" reason: I don't know (False Core Distractor). It also can become a *way* to overcome the False Core driver so that the False Self Compensator (re-merges) and can be obtained. In the above, "I have to know," then "I exist" (False Self), if I exist (False Self), I don't have to experience "I don't know" or "I don't exist" (False Core). Here the distractor of "I don't know" is used to not experience the False Core of "I don't exist." Then, "I have to know" gets used to overcome the "I don't exist," which is an endless process.

To Illustrate

Another illustration might be the "I am alone" False Core driver who becomes obsessive-compulsive about money (i.e., "I have to have worth" because "I have no value"). This now becomes the *imagined* reason for the unconscious False Core Driver and its pain. Rather than dealing with the False Core Driver "I am alone" it presumes that *reason*: it is because "I have no worth or value" which is the False Core "I am unworthy" (that is the *reason* why I am alone). Then it tries to have value (money), the False Self of "worthy" so that it can then connect (False Self) and not "be alone" (False Core).

Chapter XVIII

EVEN MORE COMPLICATED

The "I am alone" False Core Driver decides (reason) that it is "alone" because it "does not exist." It then decides that the best way to exist is to have "value." It then obsesses about having value so that it can *exist* (overcoming *I don't exist*), then if *I have worth* I can *exist* and *connect* and not have to be alone because *I can connect*. In this complicated fashion, an *infinite* number of possible combinations can occur.

As my first mentor, Dr. Frederick Herzberg, used to say,

People have an infinite capacity to find an infinite number of ways to make themselves infinitely miserable.

Quantum Psychology would say that the False Core Driver has an infinite capacity to find an infinite number of ways to avoid itself in any combination which re-enforces itself thus keeping the False Core Driver and your entire psychology alive and well.

QUANTUM PSYCHOLOGY PRINCIPLE:

The movements away from the False Core are all an avoidance of the False Core Driver. No matter what movement is made, you always take the False Core with you, the False Core Driver drives it. The False Core Driver drives the "I" and all that "you" call "you."

QUANTUM PSYCHOLOGY PRINCIPLE:

You cannot separate the False Core-False Self from the "I". It is the "I" and all "you" *think* you are.

THE MOVEMENTS GIVE REASONS

Any movement to another False Core Driver or False Self creates a reason and a justification (philosophy) as to why.

THE INACCURATE PERCEPTION
Examples:

The "There must be something wrong with me" False Core might say, "If only I were smart, then I would be *adequate* and *perfect*," and not have to feel there must be *something wrong with me*.

An *Unworthy* False Core might say, "*I am loveless*, that's the problem. If only I could meet a *powerful* person, then I could get love, feel worth, and not be *worthless*."

The *I cannot do* False Core might say, "*I cannot do* because *I am worthless*. If only I could be *adequate*, I could be *powerful*, have *worth* and then I could *do*, and not have to feel *I cannot do*."

An *inadequate* False Core might say, "I am *inadequate* because I am *alone*. I am *alone* because there *must be something wrong with me*. If only I could only fix this and be *perfect* then I would feel *adequate*, and not have to feel *inadequate*."

An *I don't exist* False Core might say, "*I don't exist* because I am *stupid*. That's the reason. If I could *do* and create, then I could prove I was *smart* and then I would *exist* and not have to feel *I don't exist*."

An *Incomplete* False Core might say, "I am *incomplete* because *I don't know*. If I *knew*, then I could *connect* with others and not be *alone*. Then I could have experiences and not feel *incomplete*."

A *powerless* False Core might say "I am *powerless* because I feel no love. I feel *no love* because I am *incomplete*. If only I could really be (*exist*) here, I could *connect*. Then I would feel *love*. Then I would not feel so *powerless*."

The *I am Loveless* False Core might, "I'm *loveless* because there's *nothing* inside of me. I feel this nothing because I was separated from God (mom). This was because *I don't know*. If only *I knew*, I would *exist* and then I'd feel *complete* and be able to connect. Then I could get *love* and not have to feel *loveless*."

As you will see, the False Core kites to follow are simplified. However, the critical issues remain the same.

Chapter XVIII

1. Amnesia.
2. An avoidance of the False Core driver by the False Self Compensator and distractor I-dentities.

THE ASSOCIATION TRANCE RE-VISITED

Don't be fooled by content. The grooves of the record are the same. The associational trance makes you believe 1) it is not a different situation; 2) that it is the same as the past situation; and 3) the people and situation are the same now as in the past. But the grooves (melody) sound the same even though they are in a different moment in space-time. The associational trance takes new situations and people and plays them through the same grooves of the same old song.

In a Charlie Brown cartoon, Charlie is listening to a record and Lucy comes over to him and says, "What are you listening to?" "This is the most depressing music I have ever heard," he says, "The more I hear it, the more depressed I get." "That's terrible," Lucy says. Charlie Brown replies, "Do you want to hear it again?

The following are *very very* simplified ways of looking at the continued attempt to overcome the False

The Way of the Human • The False Core and the False Self

MERGER THROUGH PERFECTION

False Core Driver:
"There must be something wrong with me."
"I am imperfect."

False Self Compensator:
"I must be perfect and prove there is nothing wrong with me."
(To merger)

Desired Destination

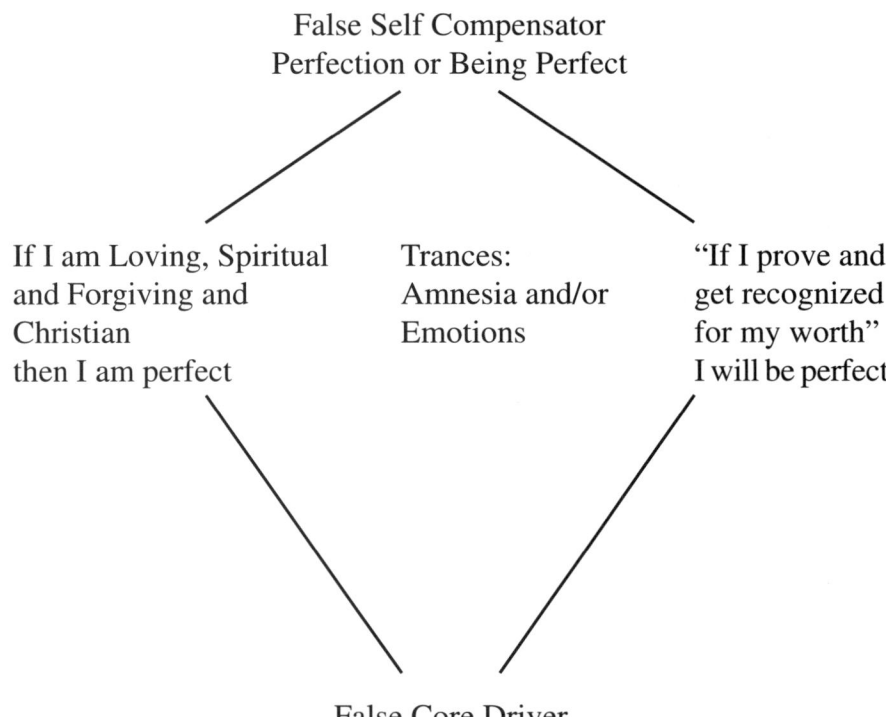

False Self Compensator
Perfection or Being Perfect

If I am Loving, Spiritual and Forgiving and Christian then I am perfect

Trances: Amnesia and/or Emotions

"If I prove and get recognized for my worth" I will be perfect

False Core Driver

Chapter XVIII

MERGER THROUGH WORTH

False Core Driver:
"I am worthless. I have no value."

False Self Compensator:
"Proving Worth and Value (to merge)."

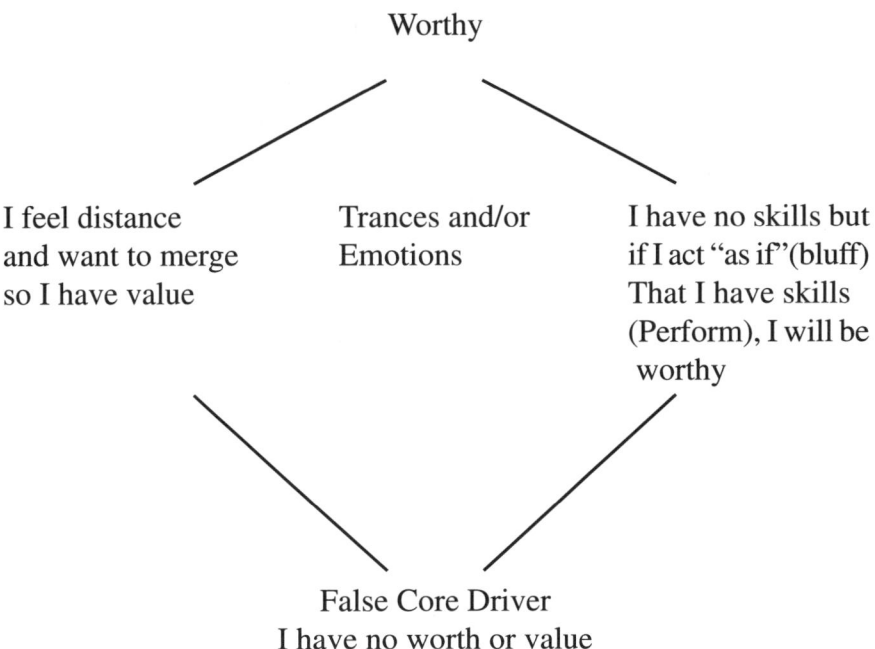

The Way of the Human • The False Core and the False Self

MERGER THROUGH DOING

False Core Driver:
"I cannot do."

False Self Compensator:
"I must over do (to merge)."

Desired Outcome

False Self Compensator
Achievement

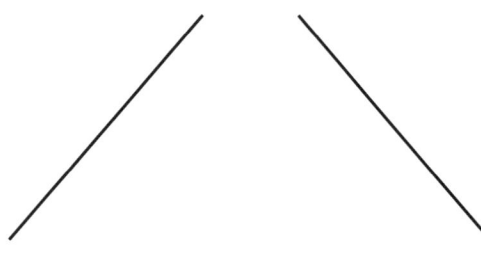

I am disorganized.
If I become organized, I can
prove worth (and merge)

Trances and/or
Emotions or
States

I am inadequate
if I prove I am
smart, then I can.
achieve.

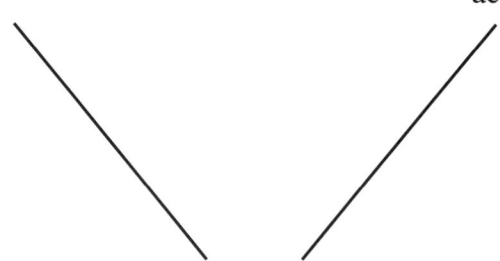

False Core Driver
"I cannot do."

Chapter XVIII

MERGER THROUGH ADEQUACY
(PROFICIENT)

False Core Driver:
"I am inadequate."

False Self Compensator:
"I must prove my proficiency (then I merge)."

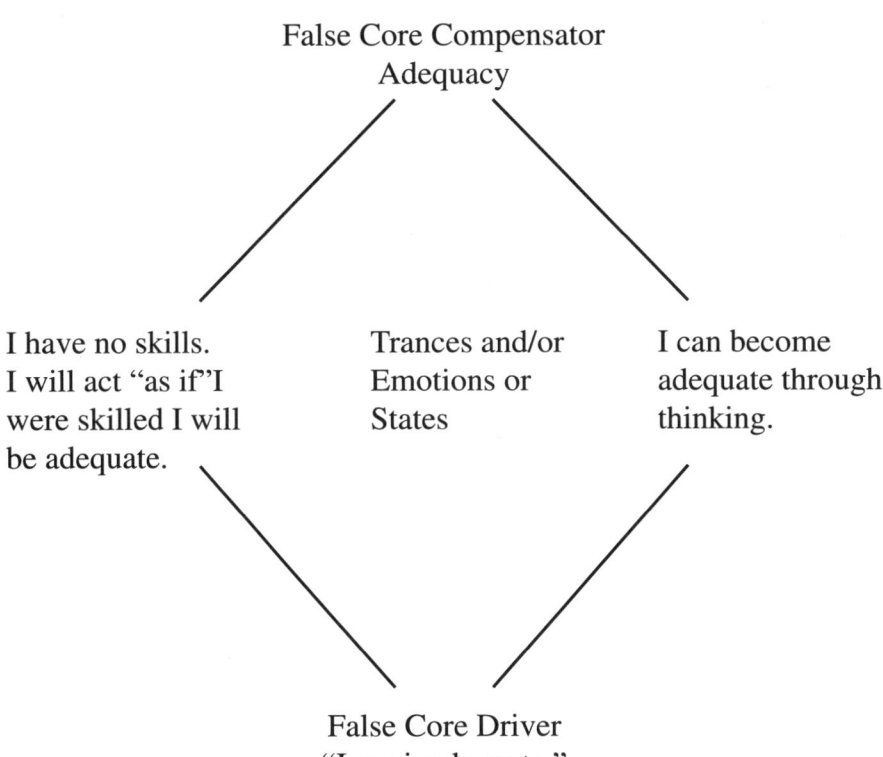

The Way of the Human • The False Core and the False Self
MERGER THROUGH EXISTENCE

False Core Driver:
"I don't exist."

False Self Compensator:
"I have to prove my existence."

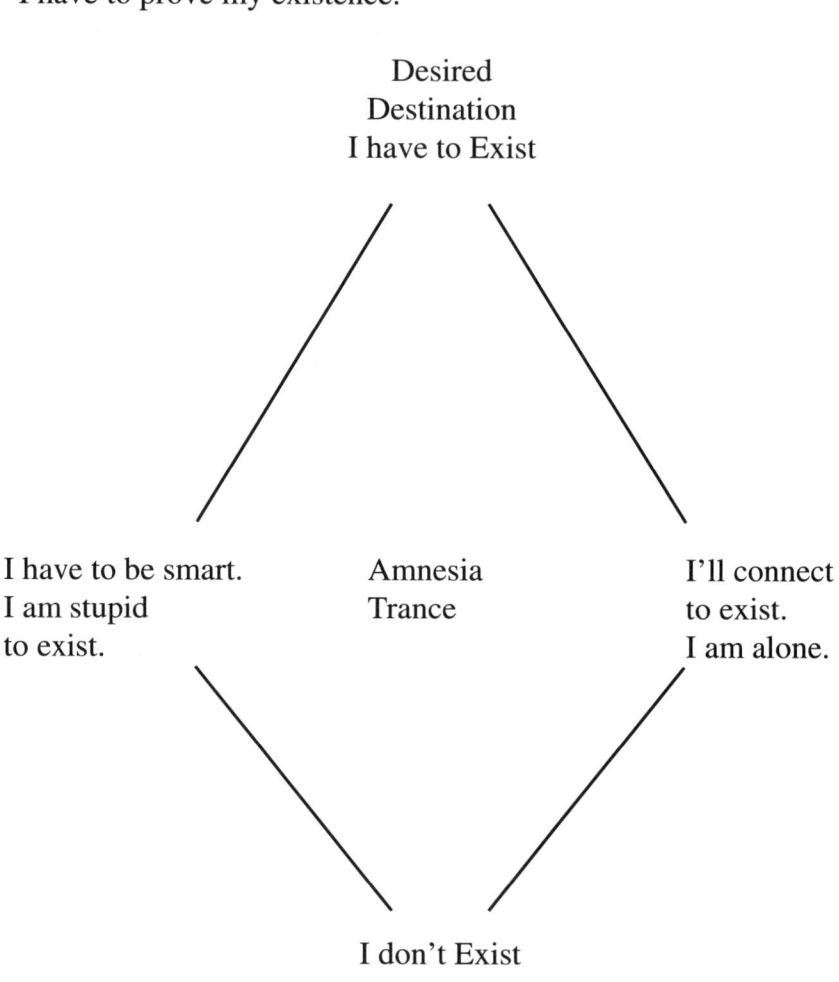

Chapter XVIII

MERGER THROUGH CONNECTION

False Core Driver:
"I am alone."

False Self Compensator:
"I have to connect."

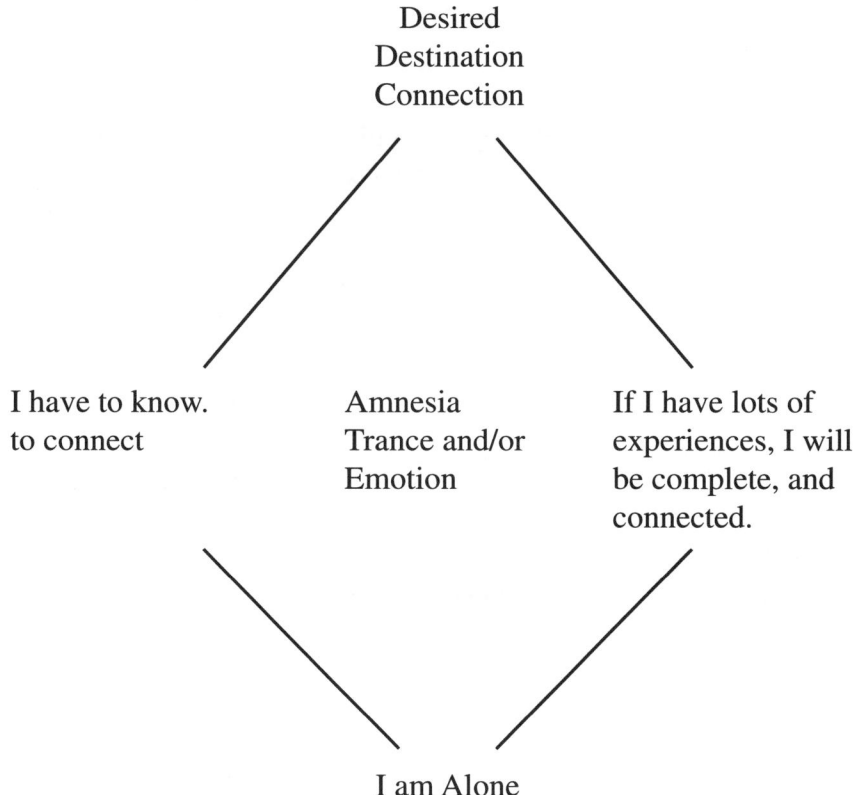

The Way of the Human • The False Core and the False Self
MERGER THROUGH BECOMING COMPLETE

False Core Driver:
"I an incomplete."

False Self Compensator:
"I must get complete or whole."

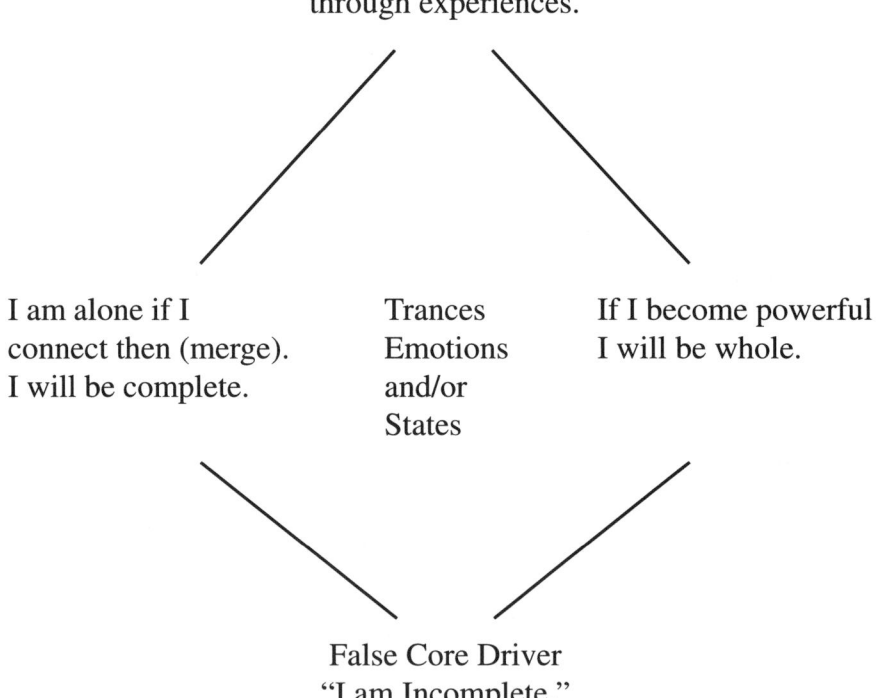

Chapter XVIII

MERGER THROUGH POWER

False Core Driver:
"I am powerless."

False Self Compensator:
"I must be powerful."

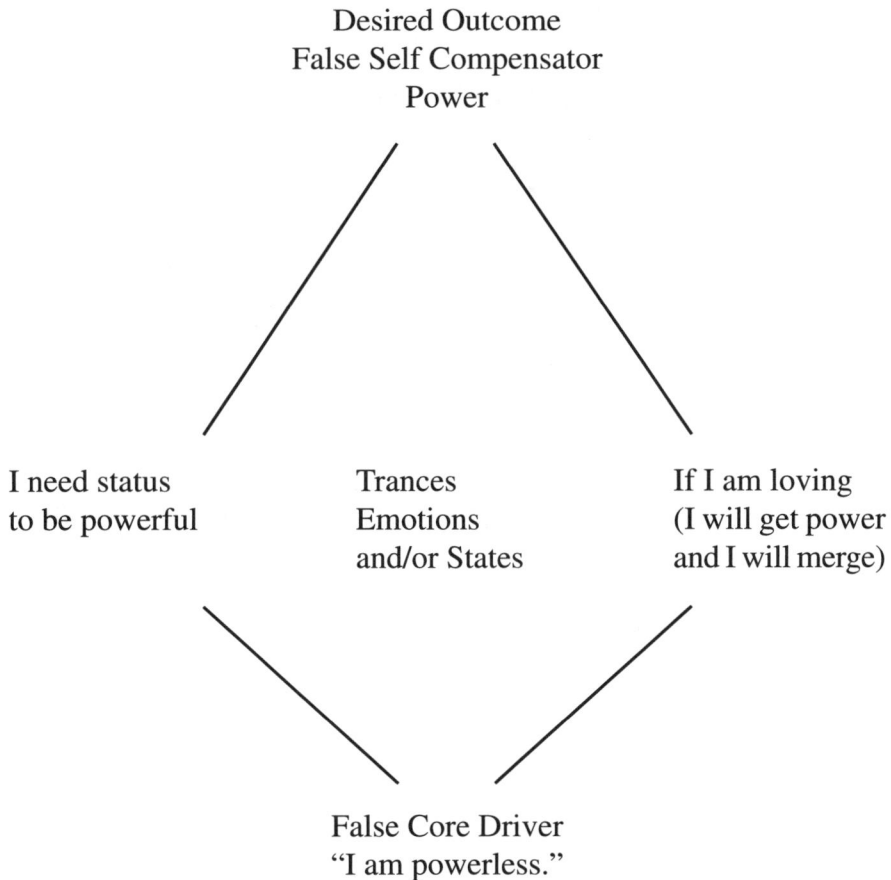

The Way of the Human • The False Core and the False Self
MERGER THROUGH LOVE

False Core Driver:
"I am loveless," "there is no love."

False Self Compensator:
"I must be loving."

Desired Outcome
False Self Compensator

"I must be loving"

If I act like a puritan, I will get love and merge

Trances
Emotions
and/or States

If I act perfect, I will get love and merge.

False Core Driver
"There is no love."

Chapter XVIII

Of course, there is more than one pattern per False Core driver. These illustrations are *very* simply possibilities. Truly, hundreds of pages could be written to describe all of the thousands of possibilities.

To discover your merger style is what is important.

EXERCISE
DIAGRAMMING AND MAPPING YOUR REACTIONS

Step I: Notice your False Core driver and place it at the bottom.
Step II: Notice another False Core distractor I-dentity or False Self you go to, to avoid your False Core.
Step III: Trace it out like a kite.
Step IV: Notice where the kite's string is fused in your body.
Step V: Notice who modeled this pattern for you.
Step VI: Take the label off of the pattern and experience it as energy.
Step VII: Allow it to go back to the (the original source) (i.e., mom/dad lineage) of the False Core-False Self.

SUMMARY

Everything you do is going to reinforce the False Core. The False Core has two distinct parts: The Driver which creates itself (like worthlessness) and the Compensator which tries to overcome itself (prove worth). False Core Drivers are self-fulfilling prophesies which are dissociated from the present-time external dimension. (They carry with them the external dimension (*memories*) from *the past*) Each False Core Driver has three main parts:

Part I: The False Core Driver
Part II: The False Self Compensator
Part III: The distractor I-dentities which provide reasons for the False Core. It is important to remember there is constant patterned movement which can: a) move to another False Core to distract; b) move to another False

Self to distract; c) move to another or any I-dentity to distract and; d) move to another emotional (state) of another False Core to distract.

QUANTUM PSYCHOLOGY PRINCIPLE:

The False Core driver must not be believed but allowed to be there *without the intention of getting rid of it*.

One student described the difference between believing, your False Core and observing it in this way.

> The other day I was sitting in *unworthy*, but it was like, *I am unworthy*. It was all day and I was depressed and it was horrible. The difference is that now I realize that there is a feeling of unworthy, but *I* wasn't unworthy. So that's the difference. If I sit in the *I am unworthy* and identify with it, then I just get depressed and I could be there anytime. But I realized that's not what you are suggesting to just feel and believe the *I am unworthy* I-dentity; but rather to be aware of it and *NOT BELIEVE IT*. Now I am observing and experiencing it as a concept rather than believing it.

Being the False Core while you are observing it is the key. If you *are* being it, do it knowingly, consciously and intentionally, using awareness, and then intentionally create it. Rather than being and creating it unknowingly, unconsciously and unintentionally.[1] Then go into the verbal **I AM** and then the non-verbal **I AM** prior to the False Core and stay there.

In the Reichian model of body-mind, energy comes through the feet and discharges through the feet. Because of the narcissistic wound the nervous system cannot process the shock of the Realization of Separation and so it freezes the energy, collapasing the levels. The energy has to go somewhere so it goes into emotions and thoughts.

---

[1] See *Trances People Live*.

Chapter XVIII

Your False Core is created as an attempt to organize the chaos of the realization of separation. At the same time, it mis-labels the **SPACIOUSNESS** of **ESSENCE-I AM** as empty (as in a lack) (and/or the belief that there is, or you have, *no core*). In this way **ESSENCE-I AM** is blamed and the shock of the Realization of Separation gets fused with the mis-labeled **SPACIOUSNESS** of **ESSENCE** as empty (as in a lack). Hence, the empty (as in a lack) is believed to mean I have *no core* and thus it is resisted. Out of the Realization of Separation, the observer I-dentity is formed. Unfortunately, this realization often becomes *spiritualized* so that rather than dealing with the separation from mom, it becomes *spiritualized* into healing, transforming and merging with God.

The False Self compensator is trying to resist the False Core. But you can also say it is a misguided attempt to heal the False Core. Most attempts fail, including spiritual attempts, because you are not dealing with the issue directly, you are using a solution based on a false conclusion.

You then fuse the False Core Driver and False Self Compensator to each other and then hook it to a particular body part or a biological inevitability. For example, eye blink, breathing, heart beat. You have to breath, your heart has to beat. These are biological. When you meditate you close your eyes and watch your breathe and/or heart beat. This slows everything down and causes a temporary loss of the False Core Driver. But when you open your eyes and breath and heart beat go back to "normal," the False Core Driver returns as well.

DISMANTLING STRUCTURES—NOT RE-ENFORCING

Nisargadetta Maharaj used to smash spiritual structures in order to get free of them. You cannot pamper I-dentities, not the spiritual ones. If you pamper them or believe in their philosophy, you only reinforce them. The False Self's philosophy will only be used as a defense to re-enforce the False Core Driver, thus the False Self drives the False Core deeper. So that's why you can't play along with them or flirt with them and hope that everything'll be okay—it never will. You have to confront these spiritual practices head on. Yogananda Paramahansa said, "The spiritual ego is the most difficult to get rid of."

The problem is that there is an odd comfort when you're in the False Core. It is a habit like smoking. Even though smoking is self-destructive, every time I smoke I feel relaxed and comfortable. The same thing with alcohol. If you like to drink, it feels great to drink a six-pack and watch TV even if it's self-destructive. In a similar way, the False Core driver is familiar. That's why it is so hard to pop.

PRIOR TO THE FALSE CORE DRIVER

The first belief structure is "I am one." Next, you experience "I am separate" because (*fill in the blank*). Then you have two structures. This merger-separation response is directly related to the nervous system and the brain. The biological drive to merge-separate is hard wired into the nervous system and is a survival response. Freud confused this merger response with sex. The libido, however, is the merger response and is the cornerstone of the False Core-False Self.

What this means is that the False Core-False Self is an aberration which comes from the natural merger (False Self)-Separation (False Core). Once the shock of the Realization of Separation occurs the nervous system gets scrambled, confusing the present with the past and projecting the past onto the future. What you call you and your psychology begins to solidify. If you watch your thoughts and fantasies they are all organized by the biological response of the merger-separation process. "I am one" is a belief. Why? Because when you are *one*, there is no subject-object, nobody there to declare or believe "I AM ONE." In other words, without using your thoughts, memory, emotions, associations or perceptions, what does *one* or *not one* mean? *One* is simply a description of the experience, it isn't the experience itself and only emerges *after* the experience itself.

Actually, "I am one" is a construct, a belief, not a realizable experience. All of your spiritual groups are driven by it. In order to experience oneness, there has to be two. It is part of the looping that keeps the False Core Driver going. I AM ONE is prior to the False

Chapter XVIII

Core and philosophizes that I am separate because (*fill in the blank*) (False Core Driver). The non-verbal **I AM** is prior to the I am One concept (see Volume III).

If you were in what went before "I am One," there would be no awareness of ONE. In this way, the *idea* or concept of being one happens simultaneously with the realization of separation.

QUANTUM PSYCHOLOGY PRINCIPLE:
The first belief is I am One. The second belief is I am separate.

Nisargadatta Maharaj, "Anything after **I AM** should be discarded.

QUANTUM PSYCHOLOGY PRINCIPLE:
Facing death brings up the narcissistic wound and the False Core more than anything else.

Spirituality is organized in the following way: I once was *one* but now I am *separate*. Different techniques are then offered to show you how to be *one* again. You light candles, you do 47 pranams (bow down), you look to the East, the West. Spiritual and psychological systems which don't understand this reinforce the entire looping process. The False Core Driver is the reason you give yourself for being *separate*, and *I am/was* is an afterthought because there are no thoughts, memory, emotions, associations or perceptions prior to the I AM ONE concept.

TRIGGER—EMPTINESS AND VULNERABILITY

What triggers the False Core? The last time you were pure awareness was prior to the shock of the Realization of Separation because there was No-I (no separate I), hence, the shock got fused with **ESSENCE-I AM** and *Vulnerability*. Who is more vulnerable than an infant? This is one of the key problems for people in relationships. It is not the fear of vulnerability or intimacy, but the fear that *if* I am vulnerable, the shock and wound will recur.

Some people feel it is better to not have intimacy, preferring to be separate than to risk re-experiencing the shock they imagine will happen. But remember that psychology comes after biology and it is biology's survival mechanism which creates your psychology with all its myriad stories and justifications as a way to prevent the vulnerability of **ESSENCE** and **I AM** which was the *last reference point* before the body experienced the shock and the I am *separate*. In this way, to avoid the intimacy and vulnerability associated with the shock we cling to the False Core.

And so, the obsessive-compulsive tendency to repeat and re-enact the False Core is a way for the body's survival mechanism to avoid the shock of the Realization of Separation and the loss of **ESSENCE-I AM** which vulnerability has come to signify.

In other words, the pure innocence, vulnerability and intimacy of an infant are *fused* with the shock. When the nervous system's *scanning-searching* device sees vulnerability whether accurate or innaccurate, the obsessive-compulsive tendency to react is automatically re-activated as a way to organize the real or imagined shock, thus bringing forth the False Core Driver, which organized the shock.

To blow out the False Core Driver, one has to go through the shock of the Realization of Separation, an act which requires enormous focus and luck. Because to blow out the False Core means to give up or go beyond the "I" called "ego" which drives your entire psychological world. This going through and beyond brings us to the entrance of "Paradise Lost"—**ESSENCE-I AM** and beyond.

To enter into the Kingdom of Heaven (**ESSENCE-I AM** and **BEYOND**), one must come as a child.

EVEN MORE THAN A FALSE CORE: A DESIRE

Suffering is caused by Desire

One of Buddha's Four Noble Truths is: All suffering is caused by desire. Quantum Psychology would say that all suffering is caused

Chapter XVIII

by the False Core-False Self desiring, substituting and re-enacting its pain.

The False Core is a self-fulfilling mechanism which continually reinforces itself.

Within your body you might experience this as a wanting desire, a pull to do or not do, act or not act, etc. This mechanism is *not* you. However, *the False Core has a desire in order to survive* and to do this it must *always* re-create and re-enforce itself.

"I" use the word "*desire*" because the False Core is part of the nervous system and, hence, it is a survival mechanism which organizes the chaos of the Realization of Separation, providing a reason so that I do not have to experience it again. Thus, False Core mechanisms which seem to be re-created as a desire or pull toward are actually a resistance against their own death. *Either knowingly or unknowingly the False Core desires and re-enforces itself.*

1. I *desire* or want to *prove* there is something wrong with me.
2. I *desire* or want to *prove* I am worthless.
3. I *desire* or want to *prove* I cannot do.
4. I *desire* or want to *prove* I am inadequate.
5. I *desire* or want to *prove* I do not exist.
6. I *desire* or want to *prove* I am alone.
7. I *desire* or want to *prove* I am incomplete.
8. I *desire* or want to *prove* I am powerless.
9. I *desire* or want to *prove* I am loveless.
10. I *desire* or want to *prove* I am not safe.
11. I *desire* or want to *prove* I am crazy.
12. I *desire* or want to *prove* I am out of control.

This re-enforcing, self-fulfilling capacity of the False Core driver always wins. No matter what you do, it is re-enacted and re-enforced; it is your *personal* "I"-dentity, experienced as an "internal desire." Forever working, like a machine or, better yet, a wind-up doll. In most situations unless you are extremely lucky the psychological or spiritual practice "you" imagine you are choosing, "it" (the False Core-False Self) is choosing. Its dying is equivalent to your dying. This subjective desire gives even more meaning to Buddha's first noble truth that "All suffering is caused by desire."

Quantum Psychology would say: "Suffering can be caused by the False Core which can be experienced as a subjective experience of desire."

PSYCHO-SPIRITUAL AND BIOLOGICAL DEFENSES

Psycho-spiritual, *and* biological "helping" systems can often be used as a defense or a re-enactment of the False Core Driver-False Self Compensator. They unknowingly support the False Self Compensator, and hence, keep the cycle (loop) alive and repeating itself.

To Illustrate

EXERCISE

Step I: Choose a spiritual, psychological or body system and philosophy you "took on."

Step II: Notice where in your body you feel your False Core Driver.

Step III: Notice the promises and guarantees of the system you belived in.

Step IV: Notice how it is an attempt to overcome the False Core Driver by re-enforcing the False Self Compensator.

Step V: Be prior to the False Core Driver and notice the connecting associational chains to the spiritual system (False Self Compensator) which is being used to reinforce the False Core driver.

Step VI: Appreciate that the False Self and system are acting as an attempt to overcome the False Core which is not you.

Step VII: Notice the size and shape of the False Core driver, the system and the **BIG EMPTINESS** they are floating in.

Step VIII: See the False Core-False Self system and the **BIG EMPTINESS** as made of the same substance.

Step IX: Turn your attention around notice what, if anything, did this process?

Chapter XVIII
QUANTUM PSYCHOLOGY AND COUPLES THERAPY

Quantum Psychology has not as of yet created a couples therapy. Recently, in several workshops I did three-hour sessions with couples and several critical areas emerged:

COUNTER TRANCE-FERENCE—TRANCE-FERENCE:

1. Seeing the partner through their father's I-dentity.
2. Seeing the partner out of their mother's I-dentity.
3. Seeing their partner from the eyes of a child and seeing the partner as mom or dad.
4. Seeing themselves as mom or dad and their partner as a child.
5. Seeing themselves through the eyes of mom or dad and imagining their partner sees them that way.
6. Seeing themselves as a child and their partner as a child.

THE FALSE CORE-FALSE SELF

7. Seeing their partner through the False Core and projecting the False Self on them.
8. Seeing their partner through the False Core and wanting them to go into their False Core too (misery loves company).
9. Seeing their partner through their False Self and seeing their partner as their False Core.
10. Seeing themselves through their False Core and seeing their partner as their compensating False Self.
11. Believing and re-enforcing their partner's False Core.
12. Believing and re-enforcing their partner's False Self.

What is clear is that you cannot have a present-time relationship if you're not in present time or if your partner isn't in present time.

Therefore, work with 1) trance-ference first (similarities and differences between them and mom/dad, etc.); and 2) process all experiences as fused dad and mom I-dentities, and "give it all" (as energy) back to their source.

When this is completed, then and only then can you see if there even is a present-time problem or is it all just "past time" fusions, trance-ferences and counter-trance-ferences.

Obviously you cannot have a present time relationship, if the people are not in present time.

A friend of mine and his wife were having marriage problems and they went to an astrologer. She looked at their charts and said, "Two halves do not make a whole, two wholes make a whole."

APPENDIX
CLOSING REMARKS

"I asked my devotees to give up their attachments—
instead they gave up their common sense."
Ramakrishna Paramahansa

DEATH AND SURVIVAL

It is always interesting to see how much our I-dentities try to survive death. Contained within each growing, sprouting, fruit-bearing seed of a plant is also its death.

We are all like a seed of a plant. Similar to a seed we too grow, sprout, and like a seed our death is contained within it. Everything in the universe contains within itself like a seed its own destruction. Death is the most dissociated and resisted experience of all. We do not understand that destruction and death are as natural as creation and birth, or as Jim Morrison said, "Nobody gets out of here alive."

I was once asked, "Why does it take so long for I-dentities to be processed and gone beyond (to die)?" "Each experience is a seed," I said, "It has not only its own life and death but its time to be born and to die." It also has its own survival and fight/flight instincts built into it.

The Danger of Diagnosis

A few years ago after realizing the False Core-False Self dilemma, I was having an experience and I noticed how "I" wanted to relate it to my False Core-False Self. In other words, I was looking through that lens and believing what "I" saw, or that "I" did this or that was because of my False Core. It all made sense, except for one thing—when I did not believe or analyze anything through the False Core-False Self's lens there was no issue.

The *lens* was the problem NOT *me*.

A few years ago Shirley Kimmell was studying object relations. "I have to stop studying object relations," she said and I asked why? She said, "Because if I look through that lens everyone has problems and looks sick."

In Raja Yoga it is said, "When a pickpocket sees a saint, they only see his pockets."

QUANTUM PSYCHOLOGY PRINCIPLE:

When you look through a psychological system you only see psychological problems.

Swami Muktanada once said, "Change your glasses," by which he meant the lenses you are looking through.

QUANTUM PSYCHOLOGY PRINCIPLE:

Don't change the lenses you are looking through, rather when your lenses are gone—there will be No-You.

**NO FRAMES OF REFERENCE,
NO REFERENCE TO FRAME.**

Questioning, re-evaluating and leaving behind what was false is a natural evolutionary process. Actually, it is part of our nervous system's *learning* to question, discovering what is psychology's and

Appendix

spirituality's False Core and discarding it. A workshop participant once said to me, "Quantum Psychology and you seem so irreverent." "I" said I learned from Nisargadetta Maharaj to be irreverent to the concepts of the False Core-False Self and very reverent to the *underlying unity*, of which we are all made.

MAP MAKING

As mentioned earlier, any premise or assumption cannot be "the truth." Rather, it is the nervous system which selects out and omits, and hence is an abstraction of the truth. We are defining abstraction as the omitting and selecting process that the Nervous System does in order to make a map of "what is." An abstraction is a conclusion which one's nervous system draws about reality. The nervous system "naturally" omits or selects out certain (unseen) Quantum events. Once this natural omission and abstraction process occurs, a premise about reality, or a map of reality, is created.

Since the map cannot represent all of the characteristics of the territory it represents, you only get a "snapshot" of the map maker's nervous system as to "how" or "why" this or that occurred and "what" can change it. All a map shows is how the map makers organize their own world. In short, the map maker is describing his or her own map.

The map "fits," if it receives enough acceptance from the world, a "successful" system emerges based on the organizing principle (original premise) which, by definition, must be false because it is a description only of the map maker's idea of the truth.

QUANTUM PSYCHOLOGY PRINCIPLE:
If the map fits—take if OFF.

A system (of premises and beliefs) based on an original premise which is false can, as the system of beliefs continues to abstract and grow, naturally move further away from the map maker's original premise. This means future premises become even more false and yield even more falsity.

QUANTUM PSYCHOLOGY PRINCIPLE:
If you have a story, it's because attention is focused on the "why" to take you away from the experience.

QUANTUM PSYCHOLOGY PRINCIPLE:
Standards are concepts which dissociate you from your humanness.

Appendix

YOU ARE NOT THE FALSE CORE AND THE FALSE SELF

See you in Volume III

With love,
Your brother, Stephen. . . .

REFERENCES

Agneesens, C. (Forthcoming). *Fabric of Wholeness: Embodying Relational Gravity.*

Almaas, A. H. (1986). *The void.* York Beach: Samuel Weiner, Inc.

American College Dictionary. (1963). New York: Random House.

Arica Institute, Inc., The. (1989). *The Arican.* New York.

Bahirjit, B. B. (1963). *The Amritanubhava of Janadeva.* Bombay: Sirun Press.

Benbu, I. (1977). *Stalking the wild pendulum.* Rochester, Vermont: Destiny Books.

Blank, G. R., & Blank, R. *Ego psychology II: Psychoanalytic developmental psychology.* New York: Columbia University Press.

Bohm, D. (1951). *Quantum theory.* London: Constable.

Bohm, D. (1980). *Wholeness and the implicit order.* London: Ark Paperbacks.

Bohm, D. (1985). *Unfolding meaning.* London: Ark Paperbacks.

Bohm, D., & Peat, D. F. (1987)). *Science, Order and Creativity.* New York: Bantam Books.

Bollas, C. (1987). *The Shadow of the object: Psychoanalysis of the unthought known.* New York: Columbia University Press.

Bollas, C. (1989). *Furies of destine: Psychoanalysis and human idiom.* London: Free Association Books.

Bourland, D., & Johnson, P. (1991). *To be or not: An e-prime anthology.* San Francisco: International Society for General Semantics.

Buddhist Text Translation Society. (1980). *The heart sutra and commentary.* San Francisco: Buddhist Text Translation Society.

Capra, F. (1976). *The tao of physics.* New York: Bantam Books.

Edinger, E. (1992). *Ego and the archetype: Individualization and the religious function of the archetype.* Boston: Shambhalla.

Gleick, James. (1987). *Chaos.* New York: Penguin Books.

Godman, D. (1985). *The teaching of Ramana Maharishi.* Ankara, London.

Hawkins, S. (1988). *A brief history of time.* New York: Bantam Books.

Herbert, N. (1985). *Quantum reality.* New York: Anchor Press.

Horner, A. J. (1985). *Object relations and the developing ego in therapy.* Northridge, New Jersey: Jason Arunsun, Inc.

Hua, Master Tripitaka. (1980). *Shurangama sutra.* San Francisco: Buddhist Text Translation Society.

Ichazo, O. (1993). *The fourteen pillars of perfect recognition.* New York: The Arica Institute, Inc.

Isherwood, C., & Prahnavarla, Swami. (1953). *How to know God: The yoga of Patanjali.* CA: New American Library.

Johnson, S. M. (1987). *Humanizing the narcissistic style.* New York: The Arica Institute, Inc.

Johnson, S. M. (1991). *The symbiotic character.* New York/ London: W. W. Norton & Co.

Kaku, M. (1994). *Hyperspace.* New York: Anchor-Doubleday Volumes.

Kaku M. (1987). *Beyond Einstein: The cosmic quest for the theory of the universe.* New York: Bantam Volumes.

Korzybski, A. (1993). *Science and sanity.* Englewood, New Jersey: Institute for General Semantics.

Korzybski, A. (1962). *Selections from Science and Sanity.* Englewood, New Jersey: International Non-Aristotelian Library Publishing Company.

Irving J. L. (1941). *Language habits in human affairs.* England, New Jersey: International Society for General Semantics

Mahler, M. (1968). *On the human symbiosis and vicissitudes of individuation.* New York: International Universe Press.

Marshall, R. J., & Marshall, S. V. (1988). *The transference-countertransference matrix: The emotional-cognitive dialogue in psychotherapy, psychoanalysis and supervision.* New York: Columbia University Press.

Mckay, M. D. M., & Fanning, P. (1981). *Thoughts and feelings: The art of cognitive stress intervention.* Oakland, CA: Harbinger Publications.

Miller, H. (1961). *Tropic of Cancer.* New York: Grove Press.

Miller, H. (1961). *Tropic of Capricorn.* New York: Grove Press.

Muktananda, Swami. (1974). *Play of consciousness.* Ganeshpuri: Shree Gurudev Ashram.

Muktananda, Swami. (1978). *I am that: The science of hamsa.* New York: S.Y.D.A. Foundation.

Mookerjit, Ajit. (1971). *Tantra asana. A way to self-realization.* Basel, Paris, New Delhi: Ravi Kumar.

Naranjo, E. (1990). *Enneatype structures: Self analysis for the seeker.* CA: Gateways IDHHB, Inc.

Nicoll, M. (1984). *Psychological commentaries on the teaching of Gurdjieff and Ouspensky.* Vol. 1. Boulder/London: Shambhala.

Nisargadatta, Majaraj. *I am that.* 1994. Durham, NC: Acorn Press

Ouspensky, P. D. (1949). *In search of the miraculous.* New York: Harcourt, Brace and World, Inc.

Palmer, H. (1988). *The Enneagram.* CA: Harper & Row.

Peat, D. F. (1987). *The bridge between matter and mind.* New York: Bantam Books.

Peat, D. F. (1988). *Superstrings and the search for the theory of everything.* Chicago: Contemporary Books.

References

Peat, D. F., & Briggs, J. (1989). *The turbulent mirror: An illustrated guide to chaos theory & the science of wholeness.* New York: Harper & Row, 1989.

Peat, D. F. (1990). *Einstein's moon: Bell's theorem and the curious quest for quantum reality.* Chicago: Contemporary Books.

Peat, D. F. (1991). *The philosopher's stone: Chaos, synchronicity, and the hidden order of the world.* New York: Bantam Books.

Postnieks, Diana. Conversations with the author.

Reich, W. (1942). *The function of the orgasm. The discovery of the orgone.* New York: World Publishing.

Riso, D. R. (1987). *Personality types: Using the Enneagram for self-discovery.* Boston: Houghton Mifflin Company.

Riso, D. R. (1988). *Understand the Enneagram.* Massachusetts: Houghton Mifflin Company.

Riso, D. R. (1987). *Humanizing the narcissistic style.* New York/London: W.W. Norton & Co.

Shah, I. (1978). *Learning how to learn: Psychology and spirituality in the Sufi Way.* London: Octagon Press

Shah, I. (1978). *A perfumed scorpion: The way to the way.* San Francisco: Harper & Row

Shakaran, R. (1991). *The spirit of homeopathy.* Bombay: Homeopathic Medical Publishers.

Singh, J. (1963). *Pratyabhijnahrdeyam: The secret of self recognition.* Delhi: Motilal Banarsidass.

Singh, J. (1979). *Siva Sutra, the yoga of Supreme Identity.* Delhi: Motilal Banarsidass.

Singh, J. (1979). *Vijnanabhairava or diving consciousness*. Delhi: Motilal Banarsidass.

Singh, J. (1980). *Spanda Karikas*. Delhi: Motilal Banarsidass.

Suzuki, S. *Zen mind, beginner's mind*. New York: Weatherhill, 1970.

Talbot, M. (1981). *Mysticism and the new physics*. New York: Bantam Books.

Talbot, M. (1987). *Beyond the quantum*. New York: Bantam Books.

Talbot, M. (1991). *The holographic universe*. New York: Harper Collins.

Vithoukas, G. (1980). *The science of homeopathy*. New York: Grove Press.

Weinberg, H. L. (1959). *Levels of knowing and existence: Studies in general semantics*. Englewood, New Jersey: Institute of General Semantics.

Wolinsky, S. H. (1993). *The dark side of the inner child*. Norfolk, CT: Bramble Co.

Wolinsky, S. H. (1991). *Trances people live: Healing approaches to quantum psychology*. Norfolk, CT: Bramble Co.

Wolinsky, S. H. (1993). *Quantum consciousness*. Norfolk, Connecticut: Bramble Books.

Wolinsky, S. H. (1994). *The tao of chaos: Quantum consciousness*. Vol. II. Norfolk, CT: Bramble Books.

Wolinsky, S. H. (1995). *Hearts on Fire* Capitola, CA